# IoT Applications Computing

*Edited by Ishwar Singh, Zhen Gao and Carmine Massarelli*

Published in London, United Kingdom

IntechOpen

*Supporting open minds since 2005*

IoT Applications Computing
http://dx.doi.org/10.5772/intechopen.87800
Edited by Ishwar Singh, Zhen Gao and Carmine Massarelli

Contributors
Ana Lídia Corrêa da Silva Moreira, Jung Tae (Steve) Kim, R. Sivakumar, B. Prabadevi, G. Velvizhi, S. Muthuraja, S. Kathiravann, M. Biswajita, A. Madhumathi, Sivagami Ponnalagarsamy, V. Geetha, M. Pushpavalli, P. Abirami, Amitabh Mishra, P.M. Benson Mansingh, G. Sekar, T. Joby Titus, Ishwar Singh, Sean Hodgins, Anoop Gadhrri, Reiner Schmidt, Zarai Khaireddine, Adnan Cherif, Luiz Cruz Villares, Christer Johannes Bengs, Alex Mouapi, Nadir Hakem, Nahi Kandil, Carmine Massarelli, Claudia Campanale, Vito Felice Uricchio



First published in London, United Kingdom, 2022 by IntechOpen
IntechOpen is the global imprint of INTECHOPEN LIMITED, registered in England and Wales, registration number: 11086078, 5 Princes Gate Court, London, SW7 2QJ, United Kingdom
Printed in Croatia

British Library Cataloguing-in-Publication Data
A catalogue record for this book is available from the British Library

Additional hard and PDF copies can be obtained from orders@intechopen.com

IoT Applications Computing
Edited by Ishwar Singh, Zhen Gao and Carmine Massarelli
p. cm.
Print ISBN 978-1-83968-689-4
Online ISBN 978-1-83968-690-0
eBook (PDF) ISBN 978-1-83968-691-7

For EU product safety concerns:
IN TECH d.o.o., Prolaz Marije Krucifikse Kozulić 3, 51000 Rijeka, Croatia,
info@intechopen.com or visit our website at intechopen.com.

# Meet the editors

Ishwar Singh is an adjunct professor in the School of Engineering Practice and Technology (SEPT), McMaster University, Ontario, Canada. Previously, he was the founding associate director of the four-year BTech programs, a joint venture between McMaster University and Mohawk College, Ontario, Canada, and the chair of the process automation program. He coordinated the design, development, and implementation of process automation, automotive and vehicle technology, and biotechnology programs. He also contributed to the establishment of the energy engineering technology degree BTech program as well as online conversion of the software program. More recently, he has been involved in the establishment of the SEPT Learning Factory for Industry 4.0 Education, Training & Applied Research, and MEng programs.

Zhen Gao is an assistant professor in the School of Engineering Practice and Technology (SEPT), McMaster University, Ontario, Canada, where he is the program lead of the MEng in Systems & Technology. He has published more than 100 papers in highly reputed journals and international conferences, one book, and four book chapters in areas of advanced automation and robotics, sensors and actuators, and artificial neural networks. Dr. Gao has been serving as the associate editor for the Conference Editorial Board of the IEEE Robotics and Automation Society. He is the editor-in-chief of the *International Journal of Intelligent Machines and Robotics*. His current research interests include advanced robotics, brain-computer interfacing, artificial intelligence/ deep learning, and their applications in manufacturing and healthcare.

Dr. Carmine Massarelli is an environmental technologist at the Italian National Council of Research, Water Research Institute. He is an expert in the development of smart technologies for water management and environmental monitoring, characterization and monitoring of contaminated and degraded sites, and integration of spatial data such as standard methodologies, interoperability, and spectral data infrastructures. He is also an expert in Geographic Information Systems, database administration, programming and IT-related roles, maintenance and creation of geographic data (accuracy and quality), use of scripting, and building cartographic output applying the principles of cartography using open-source IT systems for the processing, analysis, and integration of remote sensing data with airborne and satellite sensors for thematic purposes.

# Contents

# Preface

The COVID-19 pandemic accelerated the adoption of digital and automation technologies by companies so that they can not only stay in business but also be resilient in these challenging times. Some of these technological innovations include artificial intelligence (AI), 5G and enhanced connectivity, edge computing, Internet of Behavior, additive manufacturing, quantum computing, blockchain, human augmentation, distributed cloud, augmented reality, and virtual reality. However, even the world's most technologically advanced countries, the United States included, are struggling with a gap in digital skills necessary to utilize these new technologies. A recent poll by KPMG in Canada noted that many businesses are having a hard time hiring people with the right digital skills. It is clearly a serious issue. There are two factors to consider in this regard: lack of digital skills amongst existing workforces and lack of properly trained graduates to fill digital technology jobs in commerce and industries. Many countries have launched massive initiatives to address these issues and challenges. An example of such an initiative is the Future Skills Centre programs, which partners with employers and industry leaders across Canada on projects that advance institutional and systemic change. Universities and colleges are also ramping up their efforts to address these challenges. The core digital competencies include but are not limited to electronics, coding, the Internet of Things (IoT), the Industrial Internet of Things (IIoT), Industry 4.0, and additive manufacturing. The role of computing in all these issues is still a dominant factor that needs to be considered along with political, economic, technological, and social impacts of skills' development for the future. This book provides examples of technological innovations that are rooted in IoT technologies, where sensor data, communications, and computing are integrated. The book begins with three chapters related to sensors and communication technologies, followed by two chapters on digital technologies for health care. The next two chapters address the environment and renewable energy, followed by a chapter on the epistemology of ambient intelligence. The last part of the book includes chapters that focus on IoT/IIoT, Industry 4.0, and AI in computing. This book uses an interdisciplinary approach to explain the complex issue of scientific and technological innovations largely based on intelligent computing.

**Ishwar Singh**
Adjunct Professor,
W. Booth School of Engineering Practice and Technology,
McMaster University,
Hamilton, ON, Canada

**Dr. Zhen Gao**
McMaster University Hamilton,
ON, Canada

**Dr. Carmine Massarelli**
Water Research Institute (Earth System Science and Environmental Technologies of the National Research Council of Italy),
Bari, Italy

Chapter 1

# Radiofrequency Energy Harvesting for Wireless Sensor Node: Design Guidelines and Current Circuits Performance

*Alex Mouapi, Nadir Hakem and Nahi Kandil*

## Abstract

Given their omnipresence, electromagnetic energy offers the most attractive and recent energy supply solutions for low consumption power devices. The most targeted application is the wireless Sensor (WS) node, which is indispensable in all computing systems. This work proposes the design guideline for harvesting radio-frequency (RF) energy using the Rectifying Antenna circuit known as rectenna. The rectenna design issues are then developed to introduce new solutions for optimizing the performance of the circuits. Note that the end-to-end efficiency analysis must incorporate both receiving antenna characteristics, rectifying diode parameters, and matching filter components. However, in most studies, only one or at most two of these aspects are treated. We then want to overcome this lack by offering a global view highlighting all the design issues for optimal RF/DC conversion efficiency. The specific case of rectennas based on patch antennas and Schottky diodes, easily integrated into the circuit boards, is considered. The results of this chapter show that although the harvestable energy levels of ambient RF waves are low, some recent designs offer solutions to take advantage of these ambient waves.

**Keywords:** Rectenna, Design issues, Efficiency, Patch Antenna, Schottky diodes, WS

## 1. Introduction

Because of their low cost, flexibility, mobility, and ease of integration, Wireless Sensors (WS) are increasingly used in most computing systems. For example, WSs are used for ubiquitous structural monitoring [1]. Besides, one of WS's current major applications is the Internet of Things, in which WSs send their data to a base station that makes it available on the internet. WS provides endless opportunities and poses formidable challenges, such as the fact that energy is limited due to its battery's small size. In the quest for solutions to extend the lifespan of WSs, a new research field has evolved in recent years, known as Energy Harvesting (EH) [2]. Thanks to EH techniques, it is now possible to envisage WSs having a lifespan limited only by the hardware that constitutes them. These WSs are now well known in the literature as EH-WS [3]. The rise of the EH-WSs is due to the joint efforts in the fields of microelectronics and micro-mechanics, which today make it possible to dispose of ultra-low consumption WSs. An EH process involves identifying a primary energy source in the WS vicinity and converting it into electrical energy directly usable by the WS. This study deals with RF energy, which is ubiquitous due to the extension of telecommunications systems [4].

The proposed techniques about EH rely fully on the nature of the used primary energy sources. The main sources are the sun, vibration, thermoelectric gradient, wind, internal light, radio-frequency energy, etc. Vibration and sun sources are most considered in the literature because they generate more significant amounts of energy compared to the WS energy requirements. Recently, with the growth of radio communication systems deployment, it has become possible to consider harvesting significant quantities of RF energy in different environments. Besides, RF sources do not rely on weather conditions, as is the case with the sun source or engine operation, as would be the case with the vibration source.

This chapter presents the design considerations of energy-independent wireless sensor nodes under the base of a radiofrequency energy harvesting process. The main objective is to analyze the end-to-end conversion chain of radiofrequency waves into DC energy to define the design issues related to this growing field of research. For each stage, it will be reviewing the principles through design equations and optimization solutions. This main goal will guide the writing of the chapter, with the following specific objectives:

- Provide a taxonomy for WSs powered by radiofrequency energy.

- Provide a classification of RF energy-harvesting techniques.

- Provide state of the art on the design of the RF-EH system.

- Recall the performance of the rectenna recently designed.

Firstly, the advantage of the RF energy source compared to the other commonly used primary energy sources is proposed in Section 2. The role of wireless sensors in the IoT and the capability of some currently marketed RF energy harvesters will be presented in Section 3. Section 4 dealt with the design issues of the RF energy harvesting systems. Each stage of the conversion chain will be analyzed, and the advantages and drawbacks of the proposed solutions will be established. Finally, Section 5 concludes this chapter.

## 2. Comparison of RF source with other primary energy sources

The primary energy sources considered for EH are vibration, sun, RF Energy, airflow, internal light, heat, and wind. Most computing systems require small and light WSs to influence the measurement environment as little as possible. Therefore, the power density metric is widely used by many researchers as a criterion for comparing the performance of micro-generators [5]. **Table 1** show some recent results obtained in the design of the various micro-generators [6]. These results indicate that the current RF micro-generators have power density comparable to sources such as airflow, heat, and indoor light.

Sources like vibrations and sunlight offer power densities 10 to 100 times higher than the RF source. However, in most studies on vibrations, to achieve these performances, it is necessary to increase the level of vibration, which is not desirable for many industrial applications, especially sensor applications. In [7], for instance, a piezoelectric micro-generator is designed to generate only $23.3\ nW$ for

| Primary sources | Power densities |
|---|---|
| Vibration | 10.8 $mW/cm^3$ |
| Heat | 0.78$mW/cm^2$ |
| Wind | 0.55 $mW/cm^3$ |
| Light (outdoor) | 100 $mW/cm^2$ |
| Light (indoor) | 100 $\mu W/cm^2$ |
| RF energy | 1 $mW/cm^2$ |
| Airflow | 1 $mW/cm^2$ |

**Table 1.**
*Comparison of the some power density of Main energy harvesting methods.*

*0.25* g acceleration at a frequency of *68 Hz*. It is also foreseeable that the popularity of piezoelectric micro-generators will be declining in future years, as several research projects today are exploring solutions to reduce significantly or attenuate the vibration of engines [8].

Regarding solar energy, the achieved performance is inherently impacted during the night. Also, for WSs being deployed in indoor environments like buildings or factories, solar energy may not be available. Wind micro-generators share this constraint. These also often involve a substantial aperture [9], which is not suited to accommodate IoT applications' size design limitations. RF Energy, which, for some frequencies, can cross materials such as water, plastic, paper, and concrete, seems to be the only alternative in several situations. This research area is now expanding because this harvestable energy (RF energy) is almost always available, offering solutions to facilitate the supply of WS located in hard-to-reach environments.

Another asset of the RF source lies in the used transducer, as it can also be exploited to exchange data between sensors wirelessly. The primary transducers used are shown in **Figure 1**. The WS for transmitting and receiving information

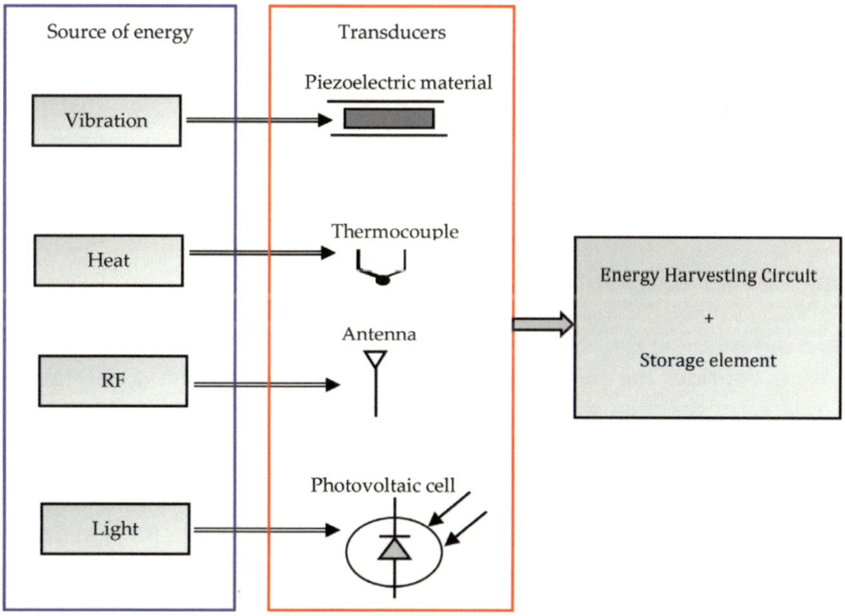


**Figure 1.**
*Some used transducers for the transformation of ambient energy.*

usually uses an antenna. The transducer for the RF source is, therefore, the same as that used by the WS to communicate. Thus, it is important for greater circuit miniaturization to consider using the RF source [10]. It is also the current trend of WS based on RF source referred to as Simultaneous Wireless Information and Power Transfer (SWIPT) technologies [11].

## 3. Autonomous wireless sensors in IoT and Mobile computing

Setting up the IoT is now possible thanks to the convenience of placing or deploying many different sensors in an environment. **Figure 2** shows the end-to-end IoT basic architecture elements [12]. The WS is the element that lays the foundation for IoT. Unlike other elements in **Figure 2** that can be placed in easily accessible locations, WSs must be able to be placed in locations such as battlefields, the deep ocean, or inhospitable terrains. Since WSs are battery-powered, it is often difficult and impossible to change or recharge their battery. Also, the WS's role in the figure below is to measure, process, and transmit data to a base station. More and more IoT applications require fast computational times, increasing the WS's energy budget. This further justifies the need for a ubiquitous charging solution such as the RF source for WS in mobile devices.

Regarding RF micro-generators, a product like the PCC110 [13] manufactured by Powercast, is a solution used to enable wireless power transmission. Its sensitivity is −17 dBm with a maximum conversion efficiency of 75%. Powercast also markets the P2110 [14], which harvests RF energy in the 915 MHz band while integrating efficient energy management solutions. This circuit can operate at incident powers below −11.5 dBm. It is also proposed in [15] the E-peas AEM40940, which offers RF energy harvesting solutions in three frequency bands 868 MHz, 915 MHz, and 2.45 GHz. This circuit offers usable DC output powers for incident RF powers between −19.5 dBm and 10 dBm. Due to the flexibility of the charging solution, these different circuits are a few examples that can be integrated into computing systems, particularly in mobile IoT applications.

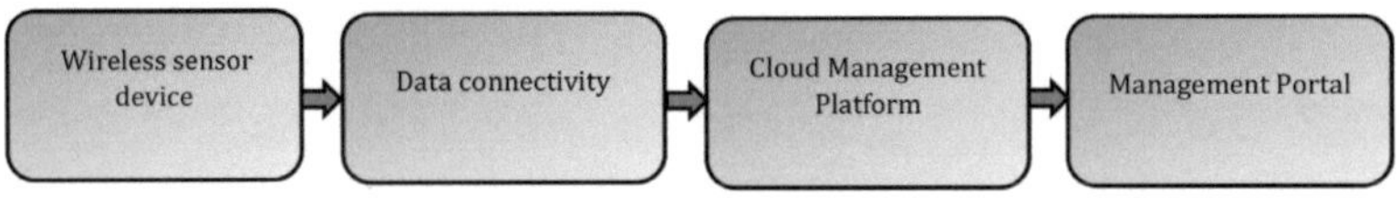


**Figure 2.**
*End-to end IoT architecture.*

## 4. Design issues of rectenna circuit

When observing at the end-to-end conversion efficiency of a rectenna, it is necessary to consider the energy propagation models, the receiving antenna, the characteristics of the rectifying diodes (RF/DC Converter), the matching filter design, and finally, the Storage Element as shown in **Figure 3**.

Energy propagation models can be used to estimate the harvestable energy levels depending on the propagation environment [16]. The receiving antenna must be designed to be optimal in the frequency band of the harvestable RF signals. The used rectifying diode must have the least loss in the targeted frequency band. The matching filter must be optimized to minimize reflection losses. A DC/DC converter is added to achieve Maximum Power Point Tracking (MPPT). Finally, when the rectenna is designed, its modeling is necessary to offer an efficient management

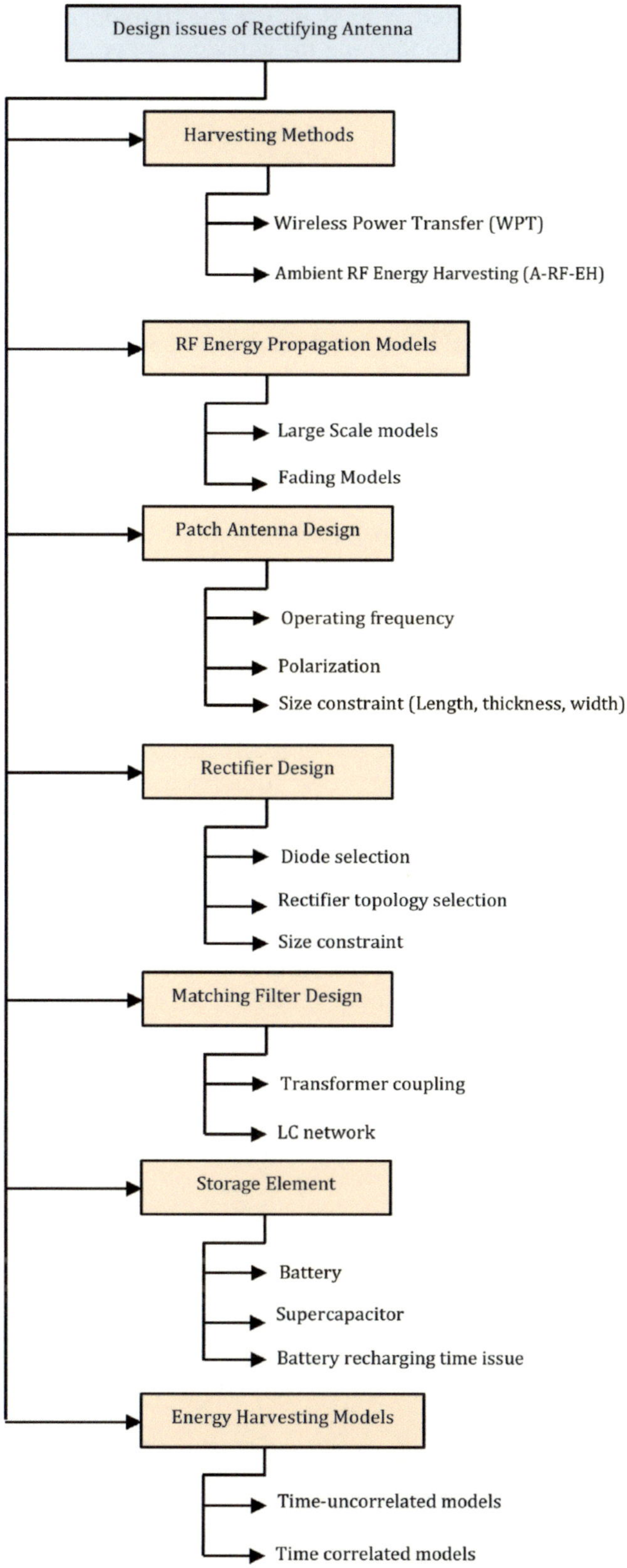


**Figure 3.**
*Outline of design issues of rectenna.*

solution for the energy harvested. Note that once the components to achieve optimal performance are selected concerning the concepts overviewed, circuit manufacturing must be addressed; this chapter does not address this issue.

### 4.1 Classification of the different techniques

When considering the use of RF energy as a power source for WSs, it is important to distinguish the Ambient RF Energy Harvesting (A-RF-EH) from Wireless Power Transfer (WPT) [17, 18] (**Figure 4**). The A-RF-EH aims to recycle energy available in the environment that comes from wireless communication devices' surrounding activity, as shown in **Figure 4(a)**. Due to potential health concerns, the environment's naturally available RF power levels are too low. However, several designers have been able to propose solutions for harvesting usable quantities of power. These solutions rely mainly on circuits' design capable of harvesting RF energy through several frequency bands simultaneously [19].

Another way to exploit RF energy is to use the WPT, as illustrated in **Figure 4(b)**. The WPT can be done either using magnetic fields to carry the electrical energy with coils or by antennas. In the case of coils, the original proposal was made by Nicolas Tesla [20] and is based on the magnetic resonance of two coils to distribute large amounts of energy to locations far from the power source. Although this concept is used by many applications such as Radio Frequency IDentification (RFID) tags [21] and biomedical devices [22], it should be mentioned that its range is limited. It would, therefore, be challenging to implement for WSs placed in hard-to-reach locations. In addition to the constrained range, the power levels are too high, bringing health issues and effects [11] when someone is close to the transmitter.

The most popular way to power the WS by RF energy is by using antennas. As opposed to the near-field application, the use of antennas is known as the far-field application. Historically, this way of transferring energy via radio waves dates to the first works of Heinrich Hertz [23]. The block diagram of the conversion of RF into DC energy via antennas is depicted in **Figure 5** [17]. A transmitting antenna sends a signal at a given power and frequency. A receiving antenna operating on the same frequency then picks up the emitted signal. An RF/DC converter is used to transform the RF signal into a DC signal. To ensure maximum power transfer between the antenna and the RF/DC converter, it is essential to use a matching circuit. The rectifier's output DC voltage is generally very low and cannot be used directly for a given application. Moreover, the value of the output DC voltage changes depending on the input RF power level. A DC-to-DC converter is thus necessary to adapt the

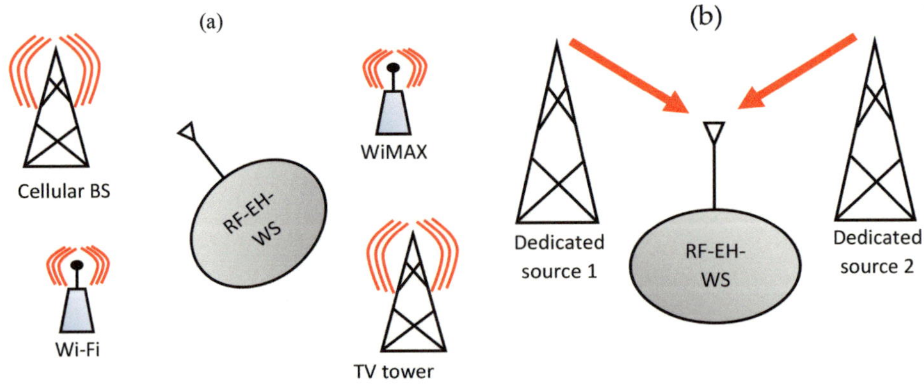


**Figure 4.**
*The technique of RF-EH. (a) A-RF-EH and (b) WPT.*

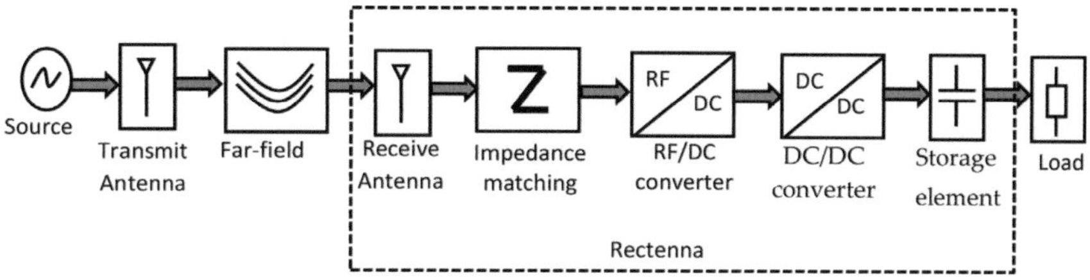


**Figure 5.**
*Generalized RF energy harvesting systems for WS.*

rectifier voltage to the storage element's voltage requirement. The combination made up of these blocks is called *rectenna* for Rectifying Antenna. The rectenna circuits' design issues then concern the channel's modeling between the transmitting and the receiving antenna, the receiving antenna's designs, the matching circuit, the rectifying circuit, the DC-to-DC converter, and finally, the choice of the storage element [24].

### 4.2 RF propagation models

The energy amount and rate received by an antenna over time are two critical parameters to be considered before the circuit design [25]. Several propagation models exist to predict the average strength of the signal received at a given distance from the transmitting antenna [16]. These models are divided into two broad kinds: large-scale and small-scale fading models.

The large-scale models are used to assess the received signal's strength over large distances between the transmitting antenna and the receiving antenna; they are then suitable for designing a WS based on the WPT. The basic model is the free space model; it is an ideal model used when the transmitting antenna and the receiving antenna have an unobstructed path. The received power is evaluated by the Friis equation as follows:

$$P_r = P_t \frac{G_t G_r \lambda^2}{(4\pi d)^2} \tag{1}$$

where $P_t$ is the transmitting power, $P_r$ is the power received at distance $d$, $\lambda$ is the wavelength of the transmitted signal, $G_t$ and $G_r$ represent the gains of the emitting and receiving antennas, respectively.

Practically, to evaluate the received power by an antenna, three basic mechanisms must be considered: reflection, diffraction, and scattering [16]. When considering ground reflection only, the Eq. (2) known as the two-ray model evaluates the received power as:

$$P_r = P_t G_t G_r \frac{h_t^2 h_r^2}{d^4} \tag{2}$$

where $P_r$,$P_t$,$G_t$,$G_r$ and $d$ are defined as above, $h_t$ and $h_r$ represent the heights of the transmit and receive antennas, respectively.

Considering all the factors influencing signal propagation (reflection, diffraction, and scattering) can be done through models derived from the combination of empirical and analytical methods [16], these models are widely used. The most popular is the path loss model, which defines the received power in a complex environment as follows:

$$P_r(d,n) = P_r(d_0)\left(\frac{d_0}{d}\right)^n \tag{3}$$

$d_0$ is a is a reference distance ($d_0 = 1\,m$ [16]) and $n$ is the path loss exponent. The value of $n$ always relates to the propagation environment features. $P_r(d_0)$ is the received power at the $d_0$ distance.

The current trend of WS powered by rectenna is the SWIPT, referred to as Simultaneous Wireless Information and Power Transfer [11]. The small-scale fading models are used to quantify the received power by a node, from a node close to it. The fading models allow evaluating the rapid fluctuations of the emitted signal's amplitude over a short period or for a short distance. Fading models consider the multiple versions of the emitted signal that reach the receiving antenna. If $N$ is the total number of possible equidistant multi-path components, then the instantaneous power received when a continuous signal is emitted is given by [16]:

$$p(t) = \left|\sum_{i=0}^{N-1} a_i \exp\left(\, j\theta_i(t,\tau)\right)\right|^2 \tag{4}$$

where $a_i$ and $\theta_i$ are respectively, the amplitude and phase of the $i^{th}$ received signal, and $\tau$ is the maximum delay.

The above summarizes some commonly used RF energy propagation models. Depending on the WPT or the A-RF-EH, deterministic models or stochastic models can be used, respectively. These models must be considered before circuit design because they make it possible to estimate the amount of harvestable energy.

### 4.3 The receiving antenna

Its role is to adequately capture the emitted signal with the right and high gain. However, the increase of the antenna gain goes with an increase in its dimensions through the equation:

$$G_R = \frac{4\pi A_e}{\lambda^2} \tag{5}$$

$A_e$ is the effective surface of the antenna, which is linked to its physical dimensions [16]. High-gain antennas are also obtained by favoring directional antennas over omnidirectional antennas. This has shown to be more effective in SWIPT [26].

To maximize the energy harvested by the antenna, particularly in the case of A-RF-EH, the studies report multi-band, broadband, and reconfigurable [27] antennas to overcome a lack of knowledge of the transmitting antennas' location and frequency.

Another important feature of the receiving antenna is its polarization, which must be circular to offer the possibility of keeping a constant DC output voltage even if the transmitting antenna or the rectenna [28] are rotating. The most widely used antennas are the dipole antennas, and the patch antennas. Since most applications have congestion as a design criterion, the patch antenna allows for easy integration; it is lightweight, low-cost, and widely considered in rectenna design. Also, these antennas are adapted to future 5G communication specifications [29]. The well-known structure of a patch antenna is shown in **Figure 6**.

The resonance frequency of the antenna, which must be the same as that of the transmitted signal, is related to length $L$ of the patch by [30]:

$$L = \frac{1}{2f_r\sqrt{\mu_0\varepsilon_0\varepsilon_e}} - 2\Delta L \quad (6)$$

$\mu_0$ and $\varepsilon_0$ represent the permeability and the dielectric permittivity of the vacuum respectively, $\Delta L$ is the length extension of the patch defined as:

$$\Delta L = 0.412h\frac{(\varepsilon_e + 0.3)\left(\frac{W}{h} + 0.264\right)}{(\varepsilon_e - 0.258)\left(\frac{W}{h} + 0.8\right)} \quad (7)$$

where $\varepsilon_e$ is the effective permittivity of the substrate, which is related to the relative permittivity as follow [31]:

$$\varepsilon_e = \frac{\varepsilon_r + 1}{2} + \left(\frac{\varepsilon_r - 1}{2}\right)\left(1 + \frac{12h}{W}\right)^{-\frac{1}{2}} \quad (8)$$

The thickness $h$ of the substrate shall satisfy the following condition:

$$h \leq \frac{1}{4f_r\sqrt{\mu_0\varepsilon_0(\varepsilon_r - 1)}} \quad (9)$$

The width $W$ of the patch influences the impedance of the antenna as well as its bandwidth. It is also related to the resonance frequency $f_r$ of the antenna as follows:

$$W = \frac{1}{2f_r}\sqrt{\frac{2}{\mu_0\varepsilon_0(\varepsilon_r + 1)}} \quad (10)$$

In most design strategies, the formulas (6) to (10) are used for the first sizing of the antenna, and then the optimization is done using an electromagnetic simulator. **Table 2** shows the gain capabilities for some patch antennas recently designed for rectenna applications.

### 4.4 RF/DC converter

To be able to supply the WSs with DC power, the RF power harvested by the antenna needs to be rectified. The RF/DC converter assumes this function. The rectification function can be implemented either by transistors or with Schottky diodes. Transistors are least-used because although they are more efficient at very

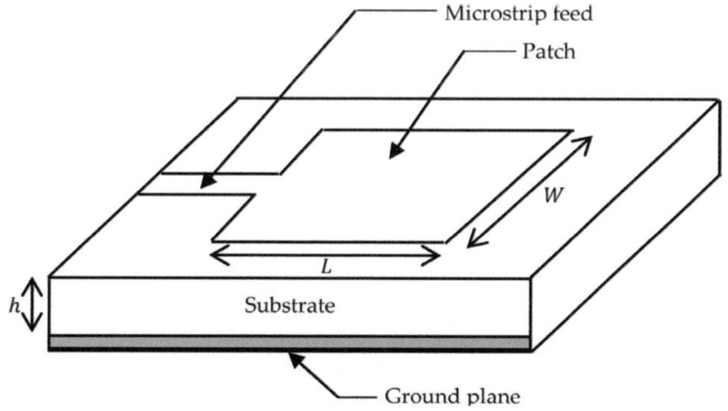


**Figure 6.**
*3D view of the rectangular patch antenna.*

| Frequency | Gain (dBi) | Sizes $L \times W \times h(mm^3)$ | Used substrate @ permittivity $\varepsilon_r$ | Ref |
|---|---|---|---|---|
| 2.36 GHz – 2.4 GHz | 6 | $60 \times 60 \times 3.2$ | RO4003@ $\varepsilon_r = 3.4$ | [30] |
| GSM 900 | 4.42 | — | FR4 @ $\varepsilon_r = 4.4$ | [32] |
| GSM 1800 | 4.32 | — | FR4 @ $\varepsilon_r = 4.4$ | [32] |
| 3G | 4.39 | — | FR4 @ $\varepsilon_r = 4.4$ | [32] |
| 2.4 GHz | 5.6 | $29 \times 37 \times 16$ | FR4 @ $\varepsilon_r = 4.6$ | [33] |
| 915 MHz | 5.9 | $90 \times 125 \times 8$ | — | [34] |
| 2.4 GHz | 7.52 | $75 \times - - \times 3.8$ | Thin Teflon @ $\varepsilon_r = 2.35$ | [31] |
| 5.5 GHz | 7.26 | $75 \times - - \times 3.8$ | Thin Teflon @ $\varepsilon_r = 2.35$ | [31] |
| GSM 900 | 7 | $115 \times 115 \times 1.6$ | RT/Duroid@ $\varepsilon_r = 2.2$ | [26] |

**Table 2.**
*Recent patch antennas gain for rectennas design.*

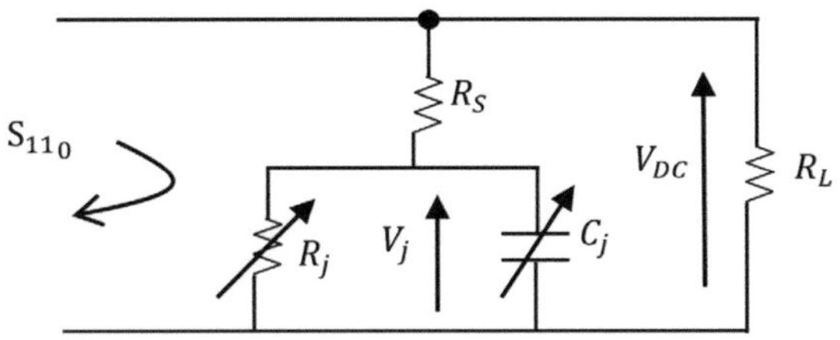


**Figure 7.**
*Small signal model of a Schottky diode.*

low levels of RF input power [35], the achieved maximum conversion efficiency remains too low compared to that obtained with Schottky diodes [24]. For this reason, the subsequent writing deals only with design issues based on Schottky diodes. The fast switching and low threshold voltage diodes are the most considered by considering the high frequencies and the low voltage level of the incident or input RF signals. The small-signal Schottky diode model shown in **Figure 7** is very often used [36].

In this model, $R_S$ is the bulk series resistance, $R_L$ is the load resistance, $R_j$ is the junction resistance, $V_{DC}$ is the voltage across the load resistance, and $C_j$ is the junction capacitance, which depends on the RF input power as follows [36]:

$$C_j = C_{j0}\sqrt{\frac{V_j}{V_j + V_{\text{DC}}}} \tag{11}$$

with $C_{j0}$, the zero-bias junction capacitance of the Schottky diode.

The leading manufacturers of the commonly used diodes are Avago, Skyworks, and Macon. **Table 3** gives the characteristics of some of the Schottky diodes most considered in the design of RF/DC converters.

In the previous subsection devoted to the receiving antenna, it was mentioned that patch antennas, being compact, lightweight, and low-cost, are the most suitable for the real applications of WSs for which congestion is one of the design constraints. However, compared to other antennas, patch antennas are narrowband and offer lower gains. Thus, the rectifying diode's conversion efficiency has become a critical design criterion for rectenna circuits [37].

| Diodes | HSMS 2810 Avago | HSMS 2820 Avago | HSMS 2850 Avago | HSMS 2860 Avago | SMS 1546 Skyworks | SMS 7621 Skyworks | SMS 7630 Skyworks | MA4E 1317 Macon | MA4E 2054 Macon |
|---|---|---|---|---|---|---|---|---|---|
| $R_S(\Omega)$ | 10 | 6 | 25 | 5 | 4 | 12 | 20 | 4 | 11 |
| $V_j(V)$ | 0.65 | 0.65 | 0.35 | 0.65 | 0.51 | 0.51 | 0.34 | 0.7 | 0.4 |
| $C_{j0}(pF)$ | 1.1 | 0.7 | 0.18 | 0.18 | 0.38 | 0.1 | 0.14 | 0.2 | 0.13 |

**Table 3.**
*Small-signal characteristics for commonly used Schottky diodes.*

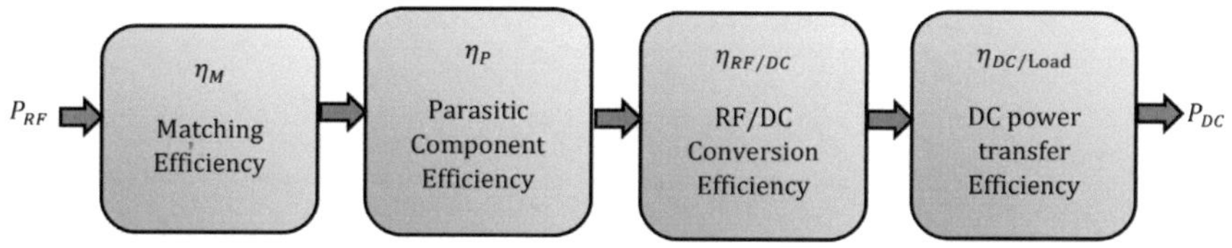


**Figure 8.**
*Efficiency link of RF/DC power conversion.*

### *4.4.1 Conversion efficiency of the rectifying diode*

When considering the transformation of the RF signal into a DC signal, the energy harvested by the antenna undergoes the four-stage losses shown in **Figure 8**. These losses are a significant factor in the choice of the rectifying diode.

The matching efficiency $\eta_M$ represents the losses due to the insertion of a matching filter between the antenna and the rectifier circuit, and it is defined in [17] by:

$$\eta_M = 1 - |S_{110}|^2 \tag{12}$$

where $S_{110}$ is the unmatched reflection coefficient defined in [38] as:

$$S_{110} = \frac{Z_D - Z_O}{Z_D + Z_O} \tag{13}$$

$Z_O$ is the output impedance of the antenna, which is generally designed to be equal to $50\ \Omega$, and $Z_D$ is the diode input impedance seen from the antenna. Depending on the internal electrical elements of the diode (Cf. **Figure 7**), $Z_D$ is expressed as follows [36]:

$$Z_D = \frac{\pi R_S}{\cos\theta_{\text{on}}\left(\frac{\theta_{\text{on}}}{\cos\theta_{\text{on}}} - \sin\theta_{\text{on}}\right) + j\omega R_S C_j\left(\frac{\pi-\theta_{\text{on}}}{\cos\theta_{\text{on}}} + \sin\theta_{\text{on}}\right)} \tag{14}$$

with $\omega = 2\pi f$, which is the pulsation of the rectenna, and $\theta_{\text{on}}$ is the diode forward-bias turn-angle. $\theta_{\text{on}}$ changes according to the incident power as follows:

$$\tan\theta_{\text{on}} - \theta_{\text{on}} = \frac{\pi R_S}{R_L\left(1 + \frac{V_j}{V_{\text{DC}}}\right)} \tag{15}$$

Considering Eqs. (13)–(15), an approximate expression of the unmatched reflection coefficient was established in [35] as follows:

$$|S_{110}|\approx\frac{R_j+(R_S-50).\left(C_j^2.R_j^2.\omega^2+1\right)}{R_j+(R_S+50).\left(C_j^2.R_j^2.\omega^2+1\right)} \tag{16}$$

From this expression, the conclusion is drawn that at high frequencies, when $C_j^2.R_j^2.\omega^2\geq 1$, it is sufficient that $R_S$ is close enough to $50\ \Omega$ to ensure a minimum reflection coefficient, and thus also to minimize the matching losses.

$\eta_P$ in **Figure 8** is the efficiency associated with parasitic losses; parasitic being undesired mechanical and electrical characteristics that limit the performance of the circuit. The parasitic component efficiency defined in [39] as:

$$\eta_P=\frac{1}{\left(1+(\omega C_j)^2R_S.R_j\right)^2} \tag{17}$$

$\eta_{RF/DC}$ in **Figure 8** is the RF/DC conversion efficiency; it is related to the elements of the diode through the following Equations [33, 36].

$$\eta_{\mathrm{RF/DC}}=\frac{1}{1+A+B+C} \tag{18}$$

with

$$\begin{cases} A=\dfrac{R_L}{\pi R_S}\left(1+\dfrac{V_j}{V_{\mathrm{DC}}}\right)^2\left[\theta_{\mathrm{on}}\left(1+\dfrac{1}{2\cos^2\theta_{\mathrm{on}}}\right)-1,5\tan\theta_{\mathrm{on}}\right] \\ B=\dfrac{R_S.R_L.C_j^2.\omega^2}{2\pi}\left(1+\dfrac{V_j}{V_{\mathrm{DC}}}\right)\left[\dfrac{\pi-\theta_{\mathrm{on}}}{\cos^2\theta_{\mathrm{on}}}+\tan\theta_{\mathrm{on}}\right] \\ C=\dfrac{R_L}{\pi R_S}\left(1+\dfrac{V_j}{V_{\mathrm{DC}}}\right)\dfrac{V_j}{V_{\mathrm{DC}}}\left[\tan\theta_{\mathrm{on}}-\theta_{\mathrm{on}}\right] \end{cases} \tag{19}$$

$\eta_{\mathrm{DC/Load}}$ in **Figure 8** represents the efficiency of DC power transfer; it is defined in [40] as follows:

$$\eta_{\mathrm{DC/Load}}=\frac{R_L}{R_L+R_T} \tag{20}$$

where $R_T=R_S+R_jI_{\mathrm{Load}}$ [41] is the Thevenin resistance seen by the load, and $R_L$ is the load resistance. To maximize $\eta_{DC/\mathrm{Load}}$, it is necessary to use a Maximum Power Point Tracking (MPPT) circuit.

Using Eqs. (12), (17), (18) and (20), a comparison of the RF/DC conversion efficiencies of the four Avago diodes, whose characteristics are reported in **Table 3**, was proposed in [33]; the obtained results are concise in **Table 4**. These results show that, for usable power levels [42], the HSMS 2850 diode is more suitable for circuit design.

### *4.4.2 Rectifier topology selection*

Once the diode is selected, it is necessary to consider the topology of the rectifier circuit. Some rectifier topologies recently used in the rectenna design are shown in **Figure 9**. The most considered are the topologies Single Series Diode (SSD), Single Parallel Diode (SPD), Full Bridge (FB), and Voltage Doubler (VD) [43]. The SSD

| $V_{DC}$(V) | Input RF power (mW) | Optimal load resistance (kΩ) | Maximum reached conversion efficiency (%) | Best rectifying diode |
|---|---|---|---|---|
| 1 | 3.52 | 0.8 | 35.5 | HSMS 2850 |
| 1.8 | 7.12 | 1.22 | 37.3 | HSMS 2850 |
| 2.5 | 14.63 | 1.13 | 37.8 | HSMS 2860 |
| 3.5 | 26.21 | 38.3 | 1.22 | HSMS 2860 |

**Table 4.**
*Best Avago rectifier diode according to the incident power level.*

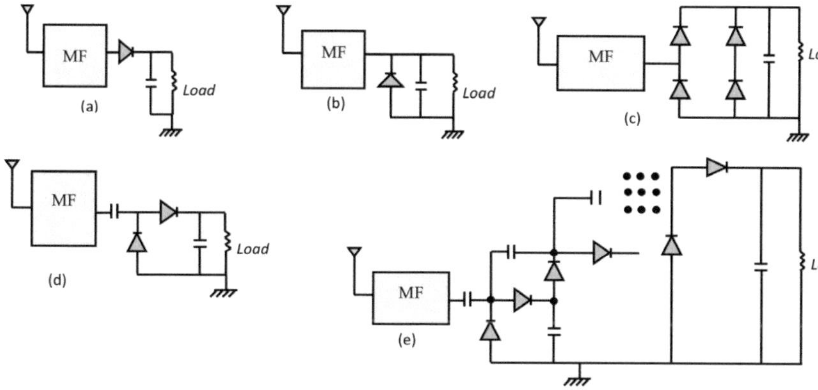


**Figure 9.**
*Most used rectifier topologies (a) SSD, (b) SPD, (c) FB, (d) VD, and (e) multistage VD.*

and SPD topologies are single-wave rectifiers, while the FB and VD topologies are full-wave rectifiers.

The characteristics of these basic topologies are proposed in **Table 5**. A comparison of the three topologies SSD, SPD, and FB, was proposed in [43] using a Rectenna Figure of Merit (RFoM) defined as follow:

$$\text{RFoM}\ (P_{\text{in}}) = V_{\text{OC}} \times \eta_{\text{optimal load}} \tag{21}$$

where $V_{\text{OC}}$ is the open-circuit voltage of the rectifier and $\eta_{\text{optimal load}}$ is the conversion efficiency reached on the optimal load of the rectifier circuit. The results obtained in [43] are that the SSD topology is best suited for the low level of input power (*−5* dBm to *0* dBm), while the SPD topology is the most efficient for medium input power level (*0* to *+15* dBm); finally, the FB topology fits better for rectennas operating at so-called high incident power levels ( > *15* dBm). However, in the literature, the most widely used topology is VD because of its voltage multiplier character [33].

It is also possible to amplify the rectified output voltage several times using several stages of voltage doubling (**Figure 9e**) [44]. When $n$ voltage doublers are set in series and connected to the load $R_L$, the output voltage across the load is expressed in [30, 45] as:

$$V_{DC} = V_{\text{OC}} \frac{nR_L}{nR_O + R_L} \tag{22}$$

where $R_L$ is the load resistance of the rectifier and $V_{\text{OC}}$ it open-circuit voltage.

| Topologies | Description | Advantages | Drawback | Applications |
|---|---|---|---|---|
| SSD | Easy to implement because it uses a single diode. | Suitable for very low power applications. | Low output DC voltage | A-RF-EH |
| SPD | Similar to SSD topology with the same performance. Instead, rectifies the negative alternation. | Suitable for very low power applications. | Low output DC voltage | A-RF-EH |
| FB | Uses the Graëtz bridge as in low-frequency power electronics. | Good conversion efficiency at high power. | Insensitive to small tensions. | WPT |
| VD | Simple Structure for rectifying the two alternations. | Higher output DC voltage. | Conversion efficiency lower than that of SSD and SPD topologies. | A-RF-EH<br>WPT |
| Multi-stage VD | Complex structure using several diodes to amplify the signal | Higher output DC voltage. | Low conversion efficiency due to losses in the diodes. | A-RF-EH<br>WPT |

**Table 5.**
*Comparison of main rectifier topologies.*

Although multi-stage VD can achieve significant voltage levels, the fact remains that they contribute to increasing the overall size of the rectenna. Also, the increase in components in the circuits contributes to an increase in losses. This is illustrated in **Figure 10**, in which up to 10 stages of voltage doublers were analyzed by simulation with Advanced Design System (ADS) software.

- **Figure 10(a)** represents the evolution of the open-circuit voltage $V_{OC}$, and it is observed that the increase in the number of stages contributes to increasing the voltage level. Saturation is observed after 4 stages.

- **Figure 10(b)** shows the conversion efficiency, and it appears that 3 stages of voltage doublers provide the best performance. Beyond that, the efficiency obtained decreases; for example, for 10 stages, maximum efficiency of less than 20% is reached at 10 dBm of incident power.

- In **Figure 10(c)**, the circuit's overall performance is analyzed according to the RFoM defined by Eq. (21). The result shows that the best compromise is reached with 4 stages.

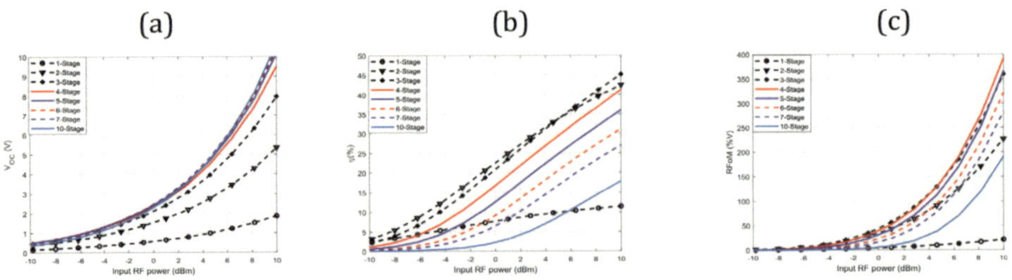


**Figure 10.**
*Mutistage VD analysis. (a) Open circuit voltage, (b) efficiency, (c) RFoM.*

### 4.5 Impedance matching

In addition to ensuring maximum power transfer between the antenna and the rectifier circuit, it also blocks the diode's harmonics. There are two main types of matching filters for rectennas: the transformer coupling and the LC network. LC networks are more popular and better suited for designing rectennas because of their ease of integration. The LC networks are made of reactive elements (coil and capacitor) which are non-dissipative [24]. The primary LC network is the low pass filter whose cutoff frequency is defined in [38] as:

$$f_0 = \frac{1}{2\pi\sqrt{LC}} \tag{23}$$

The parameter for qualitatively characterizing the adaptation is the reflection coefficient of the set consisting of the RF/DC converter and the input filter. A value of $-10\ dB$ is acceptable, according to [38], to ensure the maximum transfer of the harvested energy.

Very few analytical studies on the design of matching filters for rectenna circuits have been introduced in recent years. This is due to the power of the ADS software [33], which incorporates many tools to design and optimize the matching filter elements. In summing up the works [33, 46], the steps for creating a matching filter from the ADS software are shown in **Figure 11**.

It is shown (Cf. **Figure 11**) that from the reflection coefficient of the rectifier circuit, the ADS matching utility tool is used to generate the matching filter in a lumped component. These localized elements are the initial parameters that will then be optimized to achieve specific objectives. Three objectives are generally targeted simultaneously: the minimization of the reflection coefficient in the frequency band of interest, the maximization of the conversion efficiency, and the maximization of the DC output voltage for the expected input RF power level. The ADS software integrates several optimization techniques, the principal ones being: Hybrid, Newton, Quasi-Newton, Gradient, and Random technique. The gradient method search is the most widely used and allows for adjusting a set of variables according to an error function and its gradient. The error function usually used is the least-squares error function. Once the matching filter elements are optimized, the next step in the filter design is the transformation of the lumped component into a microstrip line. Then, the electromagnetic momentum simulator, always integrated into the ADS software, is used to predict the circuit's performance at high frequencies. This tool is used to create a physical layout to simulate the characteristics of the substrate.

Following the design steps, which are shown in **Figure 11**, it was proposed in [33], a Rectenna-based Schottky diode HSMS 2850, with a band-pass filter for an

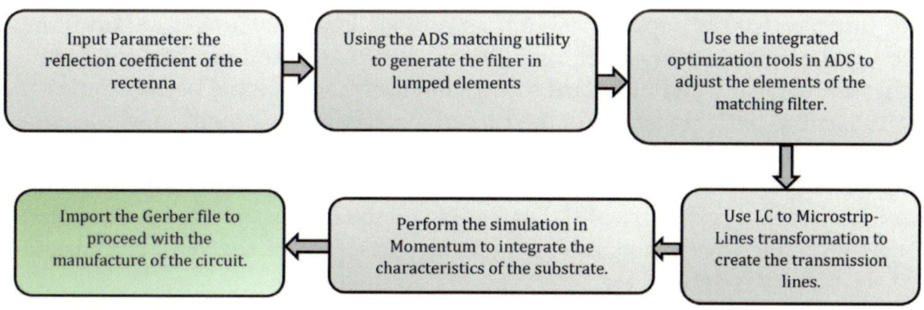


**Figure 11.**
*Design step of an optimized matching filter in ADS software.*

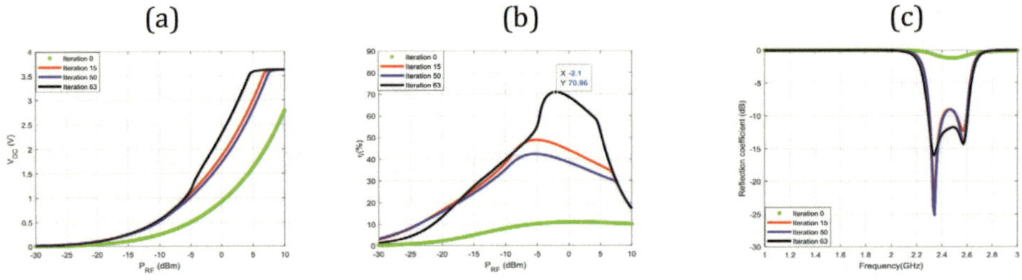


**Figure 12.**
*Optimized rectifier performance at 2.45 GHz [33]. (a) DC output voltage, (b) Conversion Efficiency, (c) reflection coefficient.*

optimal RF harvesting at *2.45 GHz*. The results of the DC output voltage, the reflection coefficient and the conversion efficiency obtained by the gradient method search are shown in **Figure 12**. The local minimum is reached after 63 iterations, and at this point, the circuit demonstrates a conversion efficiency of nearly *71%* for an incident power of $-2.1\,dBm$ $(0.61\,mW)$.

### 4.6 DC/DC converter

Most rectennas deployed in a real environment have low and variable DC output voltage due to slight fluctuations in RF input power. The voltage levels achieved cannot, therefore, directly feed the storage element. The DC / DC converter's function is then to adapt the output voltage of the rectifier to the charging voltage of the storage element. Several DC/DC converters are commercially available. For the case of rectennas design, the most suitable circuits are those with a low start-up voltage, a minimum operating power, and a high conversion efficiency over a wide range. The most appropriate circuits are then the TS3310 of TouchStone and bq25504 of Texas Instrument. A comparison of these two DC/DC converters has been proposed in [47], and it has emerged that the bq25504 converter offers better performance. However, it is less suitable for high dynamic variations of rectenna input power.

### 4.7 Storage element

Because the harvested RF energy is extremely low, it is difficult to use it to power the WS directly, hence the need for a storage element to accumulate this energy for later use. There are three main components currently used to store harvested energy: the battery, the capacitor, and the supercapacitor. Regardless of the type of the used component, the main features are capacity, voltage, energy density, power density, self-discharge, discharge depth, state of charge, and temperature effects. A comparison of the characteristics of these three components has been proposed in [48]. This study emerges that the supercapacitors can provide high power over a short time; however, the stored energy is ten times lower than that stored in a battery. This justifies the current trend of hybrid storage devices that combine both batteries and supercapacitors [49]. However, in the case of a rectenna, this solution would contribute to increasing the circuit sizes. Thus, for most rectennas involved in WSs, the energy density parameter is the most considered parameter, and it is the battery that offers the best energy density [48]. Depending on the output voltage levels of the DC/DC converter and the desired energy $E$ for the operability of the WS, the capacitance $C_b$ of the battery is defined as:

$$C_b = \frac{2E}{V_h^2 - V_\ell^2} \tag{24}$$

where $V_h$ and $V_\ell$ represent the raising threshold voltage and the falling threshold voltage of the DC/DC converter, respectively.

The general design method of WS powered by rectenna is to enslave the WS operation to the available amount of energy. Therefore, one of the major design issues is the battery recharging time, known in the literature as the duty cycle strategy. The battery recharging time knowledge helps define the duty cycle of sensors powered by the harvested energy. Depending on the battery features used, the recharging time is defined in [50] by:

$$T_r = \frac{C_b D_d V_b}{\eta P_r} \tag{25}$$

where $C_b$ is the battery capacity, $D_d$ is the discharge depth, $V_b$ is the constant operating voltage (must be chosen equal to the output rectifier DC voltage), $P_r$ is the power harvested by the receiving antenna and $\eta$ is the overall conversion efficiency of the rectenna defined as:

$$\eta = \eta_M \cdot \eta_P \cdot \eta_{RF/DC} \cdot \eta_{\text{DC/Load}} \tag{26}$$

| Input power (dBm) | Efficiency (%) | Used rectifying diode | Rectifier topology | Frequency band | Ref |
|---|---|---|---|---|---|
| −10 | 59 | HSMS 2850 | SSD | 850–950 MHz | [51] |
| 13.5 | 80.4 | HSMS 8202 | SPD | 850–950 MHz | [52] |
| −7 | 84 | 1 N6263 | FB | 850–950 MHz | [53] |
| 10 | 40* | HSMS 2850 | VD | 850–950 MHz | [54] |
| −1 | 42 | SMS 7630 | VD | 850–950 MHz | [55] |
| −2.5 | 65 | SMS 7630 | VD | 850–950 MHz | [56] |
| −20 | 20 | HSMS 2820 | SSD | 2.4–2.45 GHz | [57] |
| 10 | 66.8 | HSMS 2860 | SSD | 2.4–2.45 GHz | [58] |
| 0 | 54 | HSMS 2852 | SSD | 2.4–2.45 GHz | [59] |
| 8 | 72.8 | SMS 7630 | SSD | 2.4–2.45 GHz | [60] |
| 83 | 20 | MA4E1317 | SPD | 2.4–2.45 GHz | [61] |
| 76 | 26 | HSMS 282P | FB | 2.4–2.45 GHz | [62] |
| −13 | 9** | SMS 7630 | VD | 2.4–2.45 GHz | [55] |
| −10 | 45 | HSMS 2852 | VD | 2.4–2.45 GHz | [63] |
| 17 | 82 | MA40150–119 | SPD | 5.8 GHz | [36] |
| 17.7 | 79.5 | MA4E1317 | SPD | 5.8 GHz | [52] |
| 27 | 76 | MA4E1317 | SPD | 5.8 GHz | [64] |
| 0 | 18 | MA4E1317 | SPD | 5.8 GHz | [61] |
| 20 | 76 | MA4E1317 | SPD | 5.8 GHz | [65] |
| 0 | 54 | — | SPD | 5.8 GHz | [66] |

*is an efficiency achieved without the use of a matching filter.
**is the overall conversion efficiency considering the RF signal path losses.

**Table 6.**
*Some recent rectenna circuit performances.*

For the WS's perpetual operation, the recharging time must be equal to the time delay spent, by sensor nodes, in the sleep mode, and the energy used during the active mode must avoid draining the battery.

All the above shows that the performance of rectenna circuits depends on several parameters that have been defined in this chapter. The most considered performance criterion is the conversion efficiency of the Rectenna. A comparison of efficiency for circuits designed between 2006 and 2014 was reported in [24]. In **Table 6**, the performances of recent designs are presented. Particular attention is paid to the rectifying diode used, as well as the rectifier topology.

## 5. Conclusion

This chapter reports recent advances in the design of radiofrequency energy harvester circuits. To do this, we started by justifying the use of the RF source as a primary energy source for feeding the sensor nodes dedicated to the IoT networks. The need for completely energy-autonomous WSs in mobile computing systems has also been highlighted. We then gave an overview of the efforts carried out in the design of rectenna circuits. Current limitations due mainly to health concerns and circuit size were also mentioned. More specifically, a classification of harvesting techniques was defined, the different models of energy propagation were reviewed. The performance of the receiving patch antennas recently designed for IoT applications has been noted. The performance comparison of recently used rectifying diodes and the areas of use of the main rectifier topologies were also proposed.

## Conflict of interest

"The authors declare no conflict of interest."

## Author details

Alex Mouapi*, Nadir Hakem and Nahi Kandil
Underground Communications Research Laboratory, University of Quebec in Abitibi-Temiscamingue (UQAT), Val d'Or, Quebec, Canada

*Address all correspondence to: alex.mouapi@uqat.ca

## References

[1] N. Kurata, M. Suzuki, S. Saruwatari, and H. Morikawa, "Actual application of ubiquitous structural monitoring system using wireless sensor networks," in Proceedings of the 14th World Conference on Earthquake Engineering (14WCEE), 2008, pp. 1–9.

[2] B. Maamer, A. Boughamoura, A. M. F. El-Bab, L. A. Francis, and F. Tounsi, "A review on design improvements and techniques for mechanical energy harvesting using piezoelectric and electromagnetic schemes," *Energy Conversion and Management,* vol. 199, p. 111973, 2019.

[3] Q. Ren and G. Yao, "An energy-efficient cluster head selection scheme for energy-harvesting wireless sensor networks," *Sensors,* vol. 20, no. 1, p. 187, 2020.

[4] M. P. Aparicio, A. Bakkali, J. Pelegri-Sebastia, T. Sogorb, and V. Bou, "Radio frequency energy harvesting-sources and techniques," *Renew. Energy Util. Syst. Integr,* 2016.

[5] S. P. Beeby, M. J. Tudor, and N. White, "Energy harvesting vibration sources for microsystems applications," *Measurement science and technology,* vol. 17, no. 12, p. R175, 2006.

[6] V. C. Gungor and G. P. Hancke, "Industrial wireless sensor networks: Challenges, design principles, and technical approaches," *IEEE Transactions on industrial electronics,* vol. 56, no. 10, pp. 4258–4265, 2009.

[7] H.-C. Song *et al.*, "Ultra-low resonant piezoelectric MEMS energy harvester with high power density," *Journal of Microelectromechanical Systems,* vol. 26, no. 6, pp. 1226–1234, 2017.

[8] Q. Wang, K. Rajashekara, Y. Jia, and J. Sun, "A real-time vibration suppression strategy in electric vehicles," *IEEE Transactions on Vehicular Technology,* vol. 66, no. 9, pp. 7722–7729, 2017.

[9] S.-J. Park *et al.*, "A multi-directional wind based triboelectric generator with investigation of frequency effects," *Extreme Mechanics Letters,* vol. 19, pp. 46–53, 2018.

[10] M. Marcelić, B. Ivšić, M. Jurčević, and M. Dadić, "Estimation of energy harvesting capabilities for RF and other environmental sources," in *2018 First International Colloquium on Smart Grid Metrology (SmaGriMet)*, 2018, pp. 1–6: IEEE.

[11] T. D. P. Perera, D. N. K. Jayakody, S. K. Sharma, S. Chatzinotas, and J. Li, "Simultaneous wireless information and power transfer (SWIPT): Recent advances and future challenges," *IEEE Communications Surveys & Tutorials,* vol. 20, no. 1, pp. 264–302, 2017.

[12] Z. Sheng, C. Mahapatra, C. Zhu, and V. C. Leung, "Recent advances in industrial wireless sensor networks toward efficient management in IoT," *IEEE access,* vol. 3, pp. 622–637, 2015.

[13] S. Kim *et al.*, "Ambient RF energy-harvesting technologies for self-sustainable standalone wireless sensor platforms," *Proceedings of the IEEE,* vol. 102, no. 11, pp. 1649–1666, 2014.

[14] Powercast. (2010, 07 april). *P2110–915 MHz RF Powerharvester*. Available: https://datasheet.octopart.com/P2110-Powercast-datasheet-12525406.pdf

[15] e-peas. (2018, 07 april). *AEM40940*. Available: https://www.mouser.fr/pdf Docs/e-peas_AEM40940_DS.pdf

[16] T. S. Rappaport, *Wireless communications: principles and practice*. prentice hall PTR New Jersey, 1996.

[17] H. J. Visser and R. J. M. Vullers, "RF Energy Harvesting and Transport for Wireless Sensor Network Applications: Principles and Requirements," *Proceedings of the IEEE,* vol. 101, no. 6, pp. 1410–1423, 2013.

[18] M. Ku, W. Li, Y. Chen, and K. J. R. Liu, "Advances in Energy Harvesting Communications: Past, Present, and Future Challenges," *IEEE Communications Surveys & Tutorials,* vol. 18, no. 2, pp. 1384–1412, 2016.

[19] C. Song, P. Lu, and S. Shen, "Highly Efficient Omnidirectional Integrated Multi-Band Wireless Energy Harvesters for Compact Sensor Nodes of Internet-of-Things," *IEEE Transactions on Industrial Electronics,* 2020.

[20] N. Tesla, "Apparatus for transmitting electrical energy," ed: Google Patents, 1914.

[21] R. W. Porto, V. J. Brusamarello, I. Müller, F. L. C. Riaño, and F. R. De Sousa, "Wireless power transfer for contactless instrumentation and measurement," *IEEE Instrumentation & Measurement Magazine,* vol. 20, no. 4, pp. 49–54, 2017.

[22] P. Li and R. Bashirullah, "A wireless power interface for rechargeable battery operated medical implants," *IEEE Transactions on Circuits and Systems II: Express Briefs,* vol. 54, no. 10, pp. 912–916, 2007.

[23] W. C. Brown, "The history of power transmission by radio waves," *IEEE Transactions on microwave theory and techniques,* vol. 32, no. 9, pp. 1230–1242, 1984.

[24] X. Lu, *P. Wang*, D. Niyato, D. I. Kim, and Z. Han, "Wireless Networks With RF Energy Harvesting: A Contemporary Survey," *IEEE Communications Surveys & Tutorials,* vol. 17, no. 2, pp. 757–789, 2015.

[25] A. Mouapi, N. Hakem, and N. Kandil, "Design of 900 MHz RadioFrequency Energy Harvesting Circuit for the Internet of Things Applications," in *2020 IEEE International Conference on Environment and Electrical Engineering and 2020 IEEE Industrial and Commercial Power Systems Europe (EEEIC/I&CPS Europe)*, 2020, pp. 1–6: IEEE.

[26] I. Krikidis, S. Timotheou, S. Nikolaou, G. Zheng, D. W. K. Ng, and R. Schober, "Simultaneous wireless information and power transfer in modern communication systems," *IEEE Communications Magazine,* vol. 52, no. 11, pp. 104–110, 2014.

[27] Z. Popović *et al.*, "Scalable RF Energy Harvesting," *IEEE Transactions on Microwave Theory and Techniques,* vol. 62, no. 4, pp. 1046–1056, 2014.

[28] R. Dhara, M. Midya, M. Mitra, and S. K. Jana, "CPW-fed tetra band circular polarized antenna for wireless communication applications," in *2017 IEEE Applied Electromagnetics Conference (AEMC)*, 2017, pp. 1–2.

[29] M. Abirami, "A review of patch antenna design for 5G," in *2017 IEEE International Conference on Electrical, Instrumentation and Communication Engineering (ICEICE)*, 2017, pp. 1–3.

[30] N. Md Din, C. K. Chakrabarty, A. Bin Ismail, K. K. A. Devi, and W.-Y. Chen, "Design of RF energy harvesting system for energizing low power devices," *Progress In Electromagnetics Research,* vol. 132, pp. 49–69, 2012.

[31] W. L. Stutzman and G. A. Thiele, *Antenna theory and design*. John Wiley & Sons, 2012.

[32] M. Piñuela, P. D. Mitcheson, and S. Lucyszyn, "Ambient RF Energy Harvesting in Urban and Semi-Urban Environments," *IEEE Transactions on Microwave Theory and Techniques,* vol. 61, no. 7, pp. 2715–2726, 2013.

[33] A. Mouapi and N. Hakem, "A new approach to design autonomous wireless sensor node based on RF energy harvesting system," *Sensors,* vol. 18, no. 1, p. 133, 2018.

[34] C. Luxey, R. Staraj, G. Kossiavas, and A. Papiernik, "Antennes imprimées-Techniques et domaines d'applications," 2007.

[35] S. Hemour *et al.*, "Towards Low-Power High-Efficiency RF and Microwave Energy Harvesting," *IEEE Transactions on Microwave Theory and Techniques,* vol. 62, no. 4, pp. 965–976, 2014.

[36] J. O. McSpadden, F. Lu, and C. Kai, "Design and experiments of a high-conversion-efficiency 5.8-GHz rectenna," *IEEE Transactions on Microwave Theory and Techniques,* vol. 46, no. 12, pp. 2053–2060, 1998.

[37] D. Colaiuda, I. Ulisse, and G. Ferri, "Rectifiers' Design and Optimization for a Dual-Channel RF Energy Harvester," *Journal of Low Power Electronics and Applications,* vol. 10, no. 2, p. 11, 2020.

[38] D. M. Pozar, *Microwave engineering*. John wiley & sons, 2009.

[39] C. H. P. Lorenz, S. Hemour, and K. Wu, "Physical Mechanism and Theoretical Foundation of Ambient RF Power Harvesting Using Zero-Bias Diodes," *IEEE Transactions on Microwave Theory and Techniques,* vol. 64, no. 7, pp. 2146–2158, 2016.

[40] S. Hemour *et al.*, "Towards low-power high-efficiency RF and microwave energy harvesting," *IEEE transactions on microwave theory and techniques,* vol. 62, no. 4, pp. 965–976, 2014.

[41] S. Hemour *et al.*, "Spintronics-based devices for Microwave Power Harvesting," in *2012 IEEE/MTT-S International Microwave Symposium Digest*, 2012, pp. 1–3.

[42] P. Vecchia, R. Matthes, G. Ziegelberger, J. Lin, R. Saunders, and A. Swerdlow, "Exposure to high frequency electromagnetic fields, biological effects and health consequences (100 kHz-300 GHz)," *International Commission on Non-Ionizing Radiation Protection,* 2009.

[43] V. Marian, B. Allard, C. Vollaire, and J. Verdier, "Strategy for Microwave Energy Harvesting From Ambient Field or a Feeding Source," *IEEE Transactions on Power Electronics,* vol. 27, no. 11, pp. 4481–4491, 2012.

[44] P. Nintanavongsa, U. Muncuk, D. R. Lewis, and K. R. Chowdhury, "Design Optimization and Implementation for RF Energy Harvesting Circuits," *IEEE Journal on Emerging and Selected Topics in Circuits and Systems,* vol. 2, no. 1, pp. 24–33, 2012.

[45] K. K. A. Devi, N. M. Din, and C. K. Chakrabarty, "Optimization of the voltage doubler stages in an RF-DC convertor module for energy harvesting," *Circuits and Systems,* vol. 3, no. 03, p. 216, 2012.

[46] D. M. Pozar, "Microwave engineering 3e," *Tramsmission Lines and Waveguides,* pp. 143–149, 2005.

[47] *S. adami*, P. Proynov, B. Stark, G. Hilton, and I. Craddock, "Experimental study of RF energy transfer system in indoor environment," in *Journal of Physics: Conference Series*, 2014, vol. 557, no. 1, p. 012005: IOP Publishing.

[48] N. Rizoug, "Modélisation électrique et énergétique des supercondensateurs et méthodes de caractérisation: Application au cyclage d'un module de supercondensateurs basse tension en grande puissance," 2006.

[49] J. Liu, X. Chen, S. Cao, and H. Yang, "Overview on hybrid solar

photovoltaic-electrical energy storage technologies for power supply to buildings," *Energy conversion and management,* vol. 187, pp. 103–121, 2019.

[50] D. Altinel and G. K. Kurt, "Energy harvesting from multiple RF sources in wireless fading channels," *IEEE Transactions on Vehicular Technology,* vol. 65, no. 11, pp. 8854–8864, 2016.

[51] L. Yang, Y. J. Zhou, C. Zhang, X. M. Yang, X. Yang, and C. Tan, "Compact Multiband Wireless Energy Harvesting Based Battery-Free Body Area Networks Sensor for Mobile Healthcare," *IEEE Journal of Electromagnetics, RF and Microwaves in Medicine and Biology,* vol. 2, no. 2, pp. 109–115, 2018.

[52] J. Guo, H. Zhang, and X. Zhu, "Theoretical Analysis of RF-DC Conversion Efficiency for Class-F Rectifiers," *IEEE Transactions on Microwave Theory and Techniques,* vol. 62, no. 4, pp. 977–985, 2014.

[53] S. Ghosh, "Design and testing of RF energy harvesting module in GSM 900 band using circularly polarized antenna," in *2015 IEEE International Conference on Research in Computational Intelligence and Communication Networks (ICRCICN)*, 2015, pp. 386–389.

[54] A. Mouapi, N. Hakem, and N. Kandil, "High efficiency rectifier for RF energy harvesting in the GSM band," in *2017 IEEE International Symposium on Antennas and Propagation & USNC/URSI National Radio Science Meeting*, 2017, pp. 1617–1618.

[55] D. Masotti, A. Costanzo, M. D. Prete, and V. Rizzoli, "Genetic-based design of a tetra-band high-efficiency radio-frequency energy harvesting system," *IET Microwaves, Antennas & Propagation,* vol. 7, no. 15, pp. 1254–1263, 2013.

[56] C. Song *et al.*, "A Novel Six-Band Dual CP Rectenna Using Improved Impedance Matching Technique for Ambient RF Energy Harvesting," *IEEE Transactions on Antennas and Propagation,* vol. 64, no. 7, pp. 3160–3171, 2016.

[57] J. Zbitou, M. Latrach, and S. Toutain, "Hybrid rectenna and monolithic integrated zero-bias microwave rectifier," *IEEE Transactions on Microwave Theory and Techniques,* vol. 54, no. 1, pp. 147–152, 2006.

[58] D. Wang and R. Negra, "Design of a dual-band rectifier for wireless power transmission," in *2013 IEEE Wireless Power Transfer (WPT)*, 2013, pp. 127–130.

[59] J. A. Theeuwes, H. J. Visser, M. C. van Beurden, and G. J. Doodeman, "Efficient, compact, wireless battery design," in *2007 European Conference on Wireless Technologies*, 2007, pp. 233–236: IEEE.

[60] M. Roberg, T. Reveyrand, I. Ramos, E. A. Falkenstein, and Z. Popovic, "High-Efficiency Harmonically Terminated Diode and Transistor Rectifiers," *IEEE Transactions on Microwave Theory and Techniques,* vol. 60, no. 12, pp. 4043–4052, 2012.

[61] S. Young-Ho and C. Kai, "A high-efficiency dual-frequency rectenna for 2.45- and 5.8-GHz wireless power transmission," *IEEE Transactions on Microwave Theory and Techniques,* vol. 50, no. 7, pp. 1784–1789, 2002.

[62] S. A. Rotenberg, P. D. H. Re, S. K. Podilchak, G. Goussetis, and J. Lee, "An efficient rectifier for an RDA wireless power transmission system operating at 2.4 GHz," in *2017 XXXIInd General Assembly and Scientific Symposium of the International Union of Radio Science (URSI GASS)*, 2017, pp. 1–3: IEEE.

[63] U. Olgun, C.-C. Chen, and J. L. Volakis, "Wireless power harvesting with planar rectennas for 2.45 GHz

RFIDs," in *2010 URSI International Symposium on Electromagnetic Theory*, 2010, pp. 329–331: IEEE.

[64] R. Yu-Jiun and C. Kai, "5.8-GHz circularly polarized dual-diode rectenna and rectenna array for microwave power transmission," *IEEE Transactions on Microwave Theory and Techniques*, vol. 54, no. 4, pp. 1495–1502, 2006.

[65] H. Saghlatoon, T. Björninen, L. Sydänheimo, M. M. Tentzeris, and L. Ukkonen, "Inkjet-Printed Wideband Planar Monopole Antenna on Cardboard for RF Energy-Harvesting Applications," *IEEE Antennas and Wireless Propagation Letters*, vol. 14, pp. 325–328, 2015.

[66] K. Nishida *et al.*, "5.8 GHz high sensitivity rectenna array," in *2011 IEEE MTT-S International Microwave Workshop Series on Innovative Wireless Power Transmission: Technologies, Systems, and Applications*, 2011, pp. 19–22.

Chapter 2

# Improved Multi Target Tracking in MIMO Radar System Using New Hybrid Monte Carlo–PDAF Algorithm

*Khaireddine Zarai and Adnan Cherif*


## Abstract

This article deals with the multi-target tracking problem (MTT) in MIMO radar systems. As a result, this problem is now seen as a new technological challenge. Thus, in different tracking scenarios, measurements from sensors are usually subject to a complex data association issue. The MTT data association problem of assigning measurements-to-target or target-state-estimates becomes more complex in MIMO radar system, once the crossing target tracking scenario arises, hence the interference phenomenon may interrupt the received signal and miss the state estimation process. To avoid most of these problems, we have improved a new hybrid algorithm based on particle filter called "Monte Carlo" associated to Joint Probabilistic data Association filter (JPDAF), the whole approach named MC-JPDAF algorithm has been proposed to replace the traditional method as is known by the Extended KALMAN filter (EKF) combined with JPDAF method, such as EKF-JPDAF algorithm. The obtained experimental results showed a challenging remediation. Where, the MC-JPDAF converges towards the accurate state estimation. Thus, more efficient than EKF-JPDAF. The simulation results prove that the designed system meets the objectives set for MC-JPDA by referring to an experimental database using the MATLAB Software Development Framework.


**Keywords:** radar system, target tracking, MIMO radar, multi target tracking

## 1. Introduction

Multiple-input multiple-output (MIMO) radar system is a multistatic architecture composed of multiple transmitters and receivers, which seeks to exploit the spatial diversity of radar backscatter. In conjunction with centralized processing, MIMO radar has the potential [1] to remediate the multipath effects and improve the radar performances such as the detection, then the Multi target tracking (MTT).

In MIMO radar system, the objective of MTT is to estimate jointly at each scan the number of targets continuously moving in a given region and estimates their trajectories from noisy sensor measurements [2].

MIMO radar systems provide tracking accuracy advantages that grow proportionally with the number of transmitting and receiving radars. However, increasing the number of transmitters and receivers in MIMO radar system needs to implement new intelligent algorithms leads to increased tracking performances, these depend on the specific and intelligent tracker employed [3, 4]. Multiple target tracking (MTT) in radar system is extremely challenging, due to a lot of constraints such as the low performance of the sensor, the nature and the number of the target illuminated, the real time processing and the uncertainty of data association at that time the crossing path phenomenon is appear [5, 6], then some targets may go undetected and lead to loss their trajectories during the tracking interval.

### 1.1 Problem statement

In this paper, we concern the Motion–based Multi target tracking (MTT) problem with single sensor, which is the foundation for more complex tracking. Then, the data association problem of assigning measurements-to-target or target-state-estimates becomes more complex into MIMO radar, once the crossing target tracking phenomenon arises. Thus, the data association problem must be handled. To overpass these issues a several methods have been proposed in literature.

## 2. Related works

In order to deal with the MTT data association issues, we found in literature several methods are classified into Bayesian and other non-Bayesian filters, has been applied to address different scenarios, such as, Markov Chain Monte Carlo Data Association (MCMCDA) was proposed in [7] as a solution to replace the conventional method as known by The Multiple Hypothesis Tracking (MHT), to handle the low Signal-to-Noise Ratio (SNR) in the pre-processing phase. On the other hand, the Gaussian mixture (GM) combined with Probability Hypothesis Density (PHD), then the full GM-PHD algorithm [8] provides a promising framework to process the several measurements from multi sensors.

In [9], a joint optimization called distributed expectation-conditional maximization (DECM), has been suggested instead of the old method named Over-The Horizon Radar (OTHR) to solve the target state estimation and multipath association. Nash Equilibria method [10] is used to perform the track selection problem in MTT. The MTT by MIMO radar systems with widely distributed antennas and non-coherent processing is considered as a problem in [11], thus a hybrid algorithm is proposed based on Nearest-Neighbor Data Association (NN) and Extended KALMAN Filter (EKF).

The data association problem occurs for MTT applications and becomes more challenging in nonlinear and non-Gaussian estimation problems, hence, it is necessary to apply a Bayesian filter such as the Joint Probabilistic Data Association Filter (JPDAF) in different tracking scenarios. Thus, The JPDA algorithm calculates the association probabilities to the target being tracked for each validated measurement at the current time information, since the state and measurement equations are assumed to be linear. Therefore, in various related works we find it widely used in MTT issues, such as in [12] a new algorithm is used named Multiple Detection JPDAF (MD-PDAF) to avoid the arising multipath propagation effects for each target detection and tracking. Moreover, A Probabilistic Data Association-Feedback Particle Filter (PDA-FPF) for Multiple Target Tracking Applications is used in [13]. For multi Target tracking in passive multi-static radar system, the sequential of a multi-sensor joint probabilistic data association (S-MSJPDA) [14] has great potentials compared to the parallel architecture of a multi-sensor joint probabilistic data association (P-MSJPDA).

To avoid the data association phenomenon in MIMO radar system, our main contribution is:

- The development of a new approach based on particle filter that we called Monte Carlo – Joint probabilistic data association filter (MC-JPDAF) algorithm, to make tracking more efficient.

This paper is organized as follows; Related works in section 2. Section 3, presents our algorithm which have been used in tracking scenarios, Experimental results are discussed in sections 4, finally, the conclusion and the future works are given in section 5.

## 3. The proposed algorithm

### 3.1 Joint probabilistic data association filter (JPDAF)

JPDA algorithm aims to calculate the marginalized association probability based on all possible joint events for data association. In [12, 15], a joint event is an allocation of all measurements to all tracks. In JPDA, a feasible joint event is defined as one possible mapping of the measurements to the tracks such that: (1) each measurement (except for the dummy one) is assigned to at most one target and (2) each target is uniquely assigned to a measurement. Let $\{\theta_k = \theta_k^i\} \in \{1, 2, \dots, N_{(k/k-1)}\}$, denote the joint association event. For each pre-existed target $i \in \{1,2, \dots, N_{(k/k-1)}\}$, $\theta_k^i \in \{0,1, \dots, M_k\}$ denotes the association event, where $\theta_k^i$=j means the jth measurement is originated from the ith target and $\theta_k^i$=0 represents the dummy association in which the ith target is miss detected. JPDA assumes that each single association event is independent and the posterior of each target is:

$$P\left(X_k^i/e_k^i = 1, Z_k = \sum\nolimits_{\theta_k^i} \left(X_k^i/\theta_k^i, e_k^i = 1.Z_k\right).p\left(\theta_k^i/e_k^i = 1, Z_k\right) \tag{1}$$

### 3.2 The particle filter based on MONTE CARLO algorithm (MC)

Sequential Monte Carlo techniques are a marginal particular filter are useful for state estimation in non-linear, non-Gaussian dynamic target. These methods allow us to approximate the joint posterior distribution using sequential importance sampling.

The MC algorithm uses the sequential resampling process to avoid the filter divergence scenario during the state estimation period, particularly when using high non-linear target models and non-Gaussian distributions. Further, the process needs sufficient probability under the observed region. Accordingly, it's necessary to provide a probabilistic interpretation through the following probabilistic interpolation:

$$\mathrm{I(f) = E\ P[f(X)/Y]} = \int_1^{\mathrm{Np}} \mathrm{f(X).P(X/Y)} \tag{2}$$

### 3.3 The general MC-JPDAF algorithm

1. Initialization

   Set k = 0, generate N Samples $X^i_{t,0}$ for all targets t = 1, … ,τ indecently. $X^i_{t,0}$ Is drown from p $(X_{t,0})$, with initial weight $W^i_{t,0}=\frac{1}{N}$, for *i = 1, … , N* particles and set k = 1.

2. For i = 1, … ,N predict new particles.

$$X^{i*}_{t,k} = \mathrm{F}{*}X^i_{t,k+1} + V^i_{t,k-1}$$

3. For each particles compute the weights for all measurements (j = 0, … ,$M_k$) to targets (t = 1, … , τ) associations $W^i_{t,k}=\sum_\theta p(\theta/Z_k)$. (See Eq. (1)) And normalize the weights for each target:

$$W^i_{t,k} = \frac{w^i_{t,k}}{\sum_{i=1}^N W^i_{t,k}}$$

4. For each target, generate a new set $\{X^i_{t,k}\}_{i=1}^N$by resampling with N times from $\{X^{i*}_{t,k}\}_{i=1}^N$, where $\mathrm{P}(X^i_{t,k}=X^{i*}_{t,k})=\tilde{W}^i_{t,k}$

5. Increase k and loop

## 4. Experimental results

In this part, we attempt to prove the ability of the proposed algorithm "MONTE CARLO-JPDA" to model simulate a precise model based on target tracking parameters. This algorithm contributes to improving the state estimation of two crossing target in 2D using two separated sensors in a MIMO radar system. We will compare the results obtained from the MATLAB software.

### 4.1 Presimulation part

Firstly, we show the sensor-target geometry for tracking two crossing targets as follows (**Figure 1**):

Sensor 1:$Rx_1(0; 0\ )$; Sensor 2: $Rx_2$ (1.8e5; 0.8e5).

Initial state of the targets:

Target 1: (100e3 150; 150e3 (−10))

Target 2: (100e3 150; 148e3 10)

### 4.2 Simulation scenarios

In order to implement our algorithm, there are different variables and metrics for more accurate results interpretation were selected as follow:

- Time (T) = 200 s

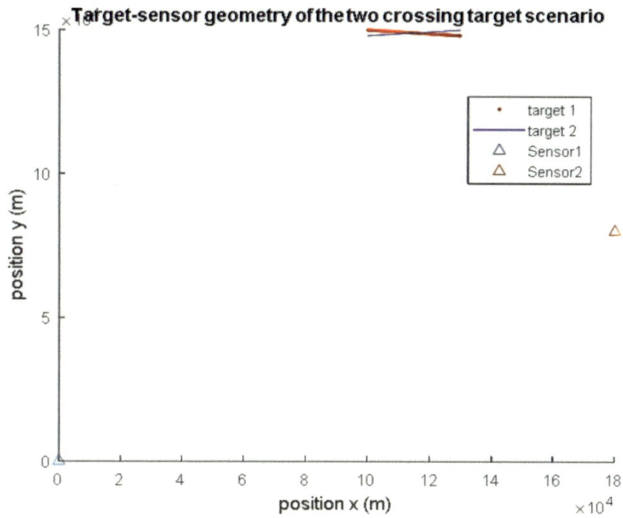


**Figure 1.**
*Initial target-sensor geometry.*

- Number of Monte Carlo simulation (MCruns) = 100 samples
- Root Mean Square error (RMSE)
- Losses Track 1 & Losses Track 2 respectively, of two crossing targets in different scenarios
- Hard assignment Simulation: using EKF-JPDAF.
- Soft assignment Simulation: using MC-JPDAF.

### *4.2.1 Two crossing targets tracking using EKF-JPDAF algorithm*

We start the tracking scenario of two crossing targets in 2-D using the conventional algorithm as known by EKF-JPDAF, the estimated trajectories and the RMSE values are given as follows (**Figures 2** and **3**):

Where: Blue dot: true target states

Green dot: estimates

Cyan star: resolved measurements

Black star: unresolved measurements

The trajectory losses of each target is given as follows:

Trajectory losses of target1: 0.187 (18.7%)

Trajectory losses of target2: 0.172 (17.2%)

According to the figures above, it is noticed that the tracking of the two targets once using EKF-JPDAF algorithm is more complex, more losses of trajectories are showed especially when the cross path phenomenon is appear, such as: Percentage of Trajectory losses for target1 is 18.7% and Percentage of Trajectory losses for target 2 is 17.2%.

### *4.2.2 Two crossing targets tracking using the suggested MC-JPDAF algorithm*

In order to improve the tracking scenario regarding the obtained results by EKF-JPDAF, we implement our new approach based on numerical filter called MC-JPDAF to

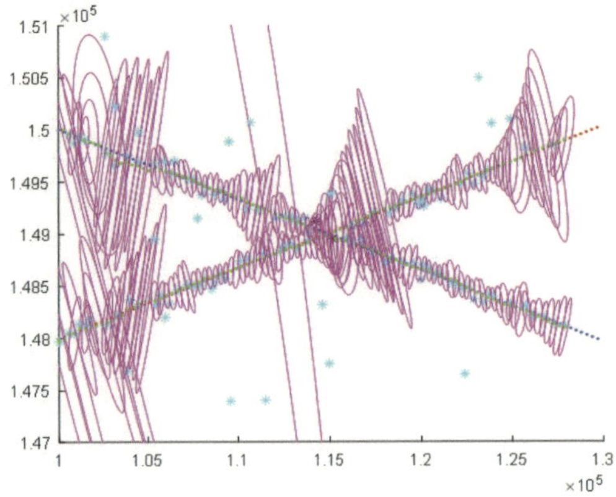


**Figure 2.**
*Trajectories of two crossing targets using measurements from the two sensors estimated by EKF-JPDAF algorithm.*

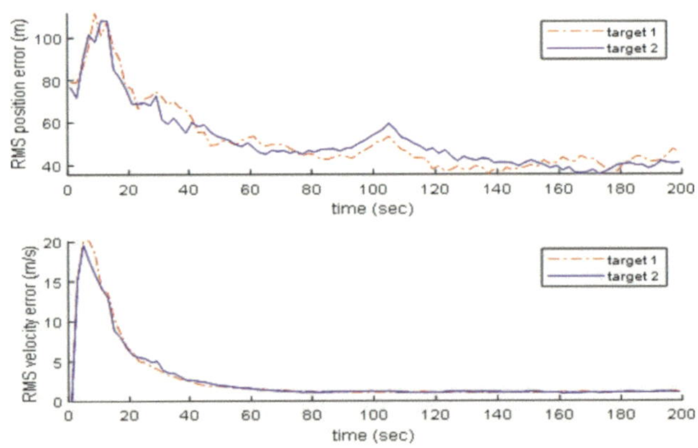


**Figure 3.**
*The RMSE position and RMSE velocity of each target.*

perform the tracking of two crossing targets in 2-D during the same estimation period (200 s). The estimated trajectories and the RMSE values are given as follows:

Where: Blue dot: true target states.

Green dot: estimates.

Cyan star: resolved measurements.

Black star: unresolved measurements.

The trajectory losses of each target is given as follows:

Trajectory losses of target1: 0, 06 (6%)

Trajectory losses of target2: 0, 07 (7%)

As shown in **Figure 4**, JPDA classifier associated to MONTE CARLO runs, provides a lower trajectories losses compared to EKF-JPDA results, such as; in **Figure 5**, during 20s the amplitude of the RMSE position is reduced from 80 m to 20 m approximately. Likewise, the RMSE velocity value goes from 22 m / s to 0.5 m / s evenhanded after 20 seconds of calculation.

The acquired results of both simulation scenarios are compared and classified in the **Table 1** hereunder.

### 4.3 Discussion

In order to strengthen the theoretical comparison in the previous section, it's clear from the results presented in **Table 1** that the EKF-JPDAF's average Ratio

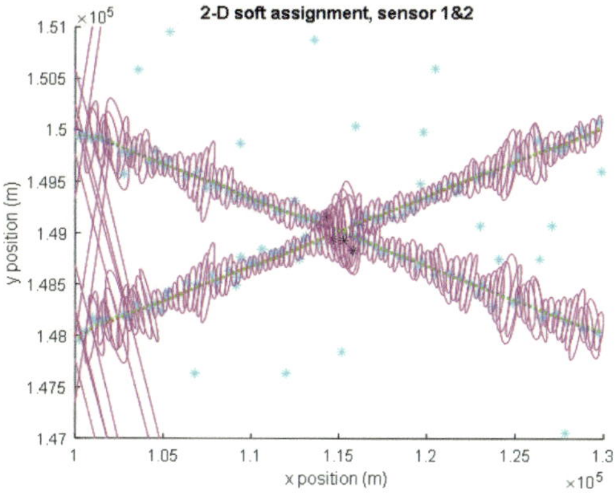


**Figure 4.**
*Trajectories of two crossing targets using measurements from both sensors estimated by MC-JPDA algorithm.*

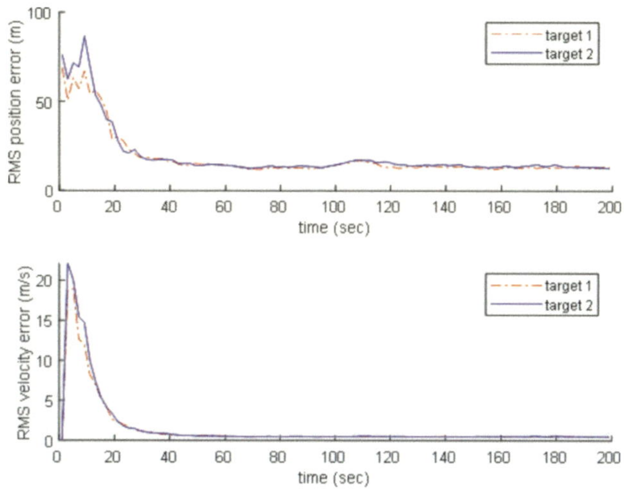


**Figure 5.**
*The RMSE position and RMSE velocity of each target.*

| | | EKF-JPDAF | MC-JPDAF |
|---|---|---|---|
| MIMO Radar 2x2 | RMSE Position (m) at T = 200 s | Target1 = 50 | Target1 = 20 |
| | | Target2 = 46 | Target2 = 20 |
| | RMSE Velocity (m/s) at T = 200 s | Target1 = 2.5 | Target1 = 0.6 |
| | | Target2 = 2.5 | Target2 = 0.6 |
| | Trajectory Losses For target 1 | 18.7% | 6% |
| | Trajectory Losses For target 2 | 17.2% | 7% |

**Table 1.**
*Comparative results.*

Mean Square Error (RMSE) is much higher than the RMSE of MC-JPDAF algorithm in both simulation scenarios. Our new MC-JPDAF method is more effective in MIMO radar system with two sensors, it gives minus tracking risk than EKF-JPDAF.

In addition to RMSE, we have added the trajectory losses percentage as a new metric for more accurate results interpretation. Thus, we notice from **Table 1** that our new hybrid algorithm have a low trajectory percentage that does not exceed a 7% of losses, which reflects the robustness of our algorithm.

The simulations are approved by comparison metrics. Therefore, in the light of this investigation, it is possible to conclude that our contribution has been verified. The new proposed hybrid MC-JPDAF algorithm estimates the state of tow crossing targets more accurately than the EKF-JPDAF algorithm. Thus it's clear to see the robustness of our approach during a long period (200 s) without performance degradation especially once the cross path phenomenon is by using a large number of Monte Carlo runs up to 100 samples.

## 5. Conclusion and future works

In conclusion, in this paper we presented a new approach to improve the MTT in MIMO Radar system as well as to avoid the filter divergence performances degradation once the crossing path phenomenon is arises.

We overcame the constraints related to the multi target tracking as mentioned in the problem statement at that point we avoided the data association issue and the filter divergence phenomenon during the tracking period. The experimental results validate what we mentioned in the theoretical part.

The MC-JPDAF approach is more efficient in complex cases which cannot be observed experimentally and even when simulated by EKF-JPDAF diverges to inappropriate results.

MC-JPDAF has a fast calculation time and converges rapidly to its related effective states. Thus, it can be used in real-time tracking.

Then, finally we have undoubtedly increased the MIMO radar system performances in MTT process by using this new approach, as a consequence we avoid the data association problem likewise the performance filter degradation. Even though having these persuasive results the method could be ameliorated by multiplying the number of targets. In our future research, we aim to implement this method aiming to enhance the multi targets tracking.

## Acknowledgements

This work has been contributed with Dr. Ibrahim BenAbdallah and the professor Cherif Adnan the editor-in-chief of the Research unit: Analysis and Signal processing of Electrical and Energetic Systems (ASPEES), department of Physics at the Faculty of sciences of Tunis- Tunisia, 2095.

## Conflict of interest

The authors declare no conflict of interest.

## Acronyms and abbreviations

| | |
|---|---|
| MTT | Multi Target Tracking |
| MC | Monte Carlo |
| PF | Particular filter |

| | |
|---|---|
| JPDAF | Joint Probabilistic Data Association Filter |
| EKF | Extended KALMAN Filter |
| MIMO | Multiple Input Multiple Output |
| MC runs | Number of Monte Carlo simulation |
| RMSE | Root Mean Square error |

## Author details

Khaireddine Zarai[1*] and Adnan Cherif[2]

1 ASPEES, Department of Physics, Faculty of Sciences of Tunis, Tunisia

2 Faculty of Sciences of Tunis, Tunisia

*Address all correspondence to: khaireddine.zarai@fst.utm.tn

## References

[1] Ankit Kumar M. Rajdip, Usha Neelakanthan. A Study on MIMO Radar Multiple Target Detection System. International Journal of Electronics Communication and Computer Technology (IJECCT). Volume 2 Issue 3, May 2012. License BY-NC-ND 4.0.

[2] Ba-ngu vo, Mahendra Mallick, Yaakov Bar-shalom, Stefano Coraluppi, Richard osborne, Ronald Mahler, and Ba-tuong vo. Multitarget Tracking. Wiley Encyclopaedia of Electrical and Electronics Engineering, Wiley. Sept. 2015. https://doi.org/10.1002/047134608X.W8275.

[3] Heyu, Gvohui Yaug,Yingsong Li and Fan Yi Meng. Design and analysis of Multiple–input Multiple Output radar system based on RF single link Technology. 10(5), 130, symmetry 2018. https://doi.org/10.3390/sym10050130.

[4] M.H. Jaward, L. Mihayolova, N. Canagarajah and D. bull. A data Association algorithm for multiple object tracking in video sequences. IEEE. 2006. DOI: 10.1049/ic: 20060565.

[5] Jean Dezan, Alberta Techamova, Pavlin Konstantinova, Erik Blasch. A Comparative Analysis of QADA-KF with JPDAF for Multi target tracking in Clutter. International conference on information Fusion. IEEEXplore: 15 August 2017. DOI: 10.23919/ICIF.2017.8009736.

[6] Cong-Thanh Do, Hoa Van Nguyen. Tracking Multiple Targets from Multistatic Doppler radar with Unknown Probability of Detection. Sensors, 2019. https://doi.org/10.3390/s19071672.

[7] Nicolas Knudde, Baptist Vandersmissen Y, Karthick Parashar, Ivo Couckuyt, Azarakhsh Jalalvand y, Andr'e Bourdoux, Wesley De Neve y, Tom Dhaene. Indoor Tracking of Multiple Persons With a 77 GHz MIMO FMCW Radar. Conference: European Radar Conference (EURAD). October 2017. DOI: 10.23919/EURAD.2017.824917.

[8] John D. Glass and Aaron D. Lanterman.MIMO Radar Target Tracking using the Probability Hypothesis Density Filter. Aerospace Conference. IEEE. 2012. DOI: 10.1109/AERO.2012.6187208.

[9] Hua Lan, Yan Liang, Zengfu Wang, Feng Yang, and Quan Pan. Distributed ECM Algorithm for OTHR Multipath Target Tracking With Unknown Ionospheric Heights. JOURNAL OF SELECTED TOPICS IN SIGNAL PROCESSING. VOL. 12, NO. 1, FEBRUARY. IEEE. 2018. DOI: 10.1109/JSTSP.2017.2787488.

[10] Nikola Bogdanovi C, Hans Driessen, Alexander Yarovoy. Track selection in Multifunction Radars: Nash and correlated equilibria. IEEE Transactions on Aerospace and Electronic Systems. 2018. DOI:10.1109/TAES.2018.2819798.

[11] Phuoc Vu1, Alexander M. Haimovich, Braham Himed. Direct Tracking of Multiple Targets in MIMO Radar. 50th Asilomar Conference on Signals, Systems and Computers. IEEE. 2016. DOI: 10.1109/ACSSC.2016.7869549.

[12] B. Habtemariam, R. Tharmarasa, T. Thayaparan, M. Mallick, And T. Kirubarajan, A Multiple-Detection Joint Probabilistic Data Association Filter', IEEE JOURNAL OF SELECTED TOPICS IN SIGNAL PROCESSING, VOL. 7, NO. 3, JUNE 2013. DOI: 10.1109/JSTSP.2013.2256772.

[13] Tao Yang, Prashant G. Mehta. Probabilistic Data Association-Feedback Particle Filter for Multiple Target

Tracking Applications. Journal of Dynamic Systems, Measurement, and Control MARCH 2018, Vol. 140. https://doi.org/10.1115/1.4037781.

[14] Xiaoyong Lyu, Jun Wang. Sequential Multi-Sensor JPDA for Target Tracking in Passive Multi-Static Radar with Range and Doppler Measurements.Volume: 7. IEEE Access. 2019. DOI: 10.1109/ACCESS.2019.2905265.

[15] Shaoming He, Hyo- Sang shin, and Antonios. Tsourdos. Joint Probabilistic Data Association Filter with unknown detection probability and clutter rate, January 2018, 18(1), 269. DOI: 10.3390/s18010269.

Chapter 3

# Vehicle Collision Avoidance System Using Li-Fi

*P.M. Benson Mansingh, G. Sekar and T. Joby Titus*


## Abstract

In recent times, large numbers of road accidents occurring all over the world are mainly due to collisions between vehicles. More than 1.2 million peoples were died in road accidents in 2019, according to the World Health Organization (WHO). Human safety features are much needed in the manufacturing of vehicles. The proposed method mainly focuses on reducing the number of accidents in our daily lives by avoiding collision between the vehicles. There are several factors corresponding to such difficult conditions that may results in death or disabilities. The causes are sudden loss of concentration of the driver, braking failure and stability issues. These criteria can be reduced only if there is a possibility for communication between the vehicles and the drivers in order to avoid accidents. There are various vehicular communication system models like Dedicated Short Range Communication and Vehicular Ad-Hoc network operating less than 5.9 GHz. These radio frequency based communication also has some limitations such as interference, congested spectrum and security. These drawbacks can be reduced by implementing the Visible Light Communication (VLC) in vehicles. It provides larger bandwidth, security, interference immunity, and high data rate. High speed data transmission and reception can be achieved using visible light based data communication system. This technology is known as Light Fidelity (Li-Fi). This chapter presents the innovative method to evade collision between two vehicles (rear and front). This communication system is cost effective with high speed data rate capabilities.


**Keywords:** Li-Fitechnology, Arduino Microcontroller, Visible light communication, Vehicle to vehicle communication, Proteus software

## 1. Introduction

Li-Fi is transmission of data through illumination by taking the fiber out of fiberoptics, sending data through an LED light bulb that varies in intensity faster than thehuman eye can follow. As this technology offers a huge bandwidth, it is unlicensed socan be used for many applications such as streaming video and music, access to theinternet, etc. The data can be transmitted through LED light which is presently observed in Electronics Devices like Hamradio, Television remotes, etc. As the number of devices connected to a particular wireless network increases the speed of the network decreases. So, in order to overcome this problem Data transmission through LED is done. Generally, data is transmitted using EM waves or else through Data Illumination.

In future data can be transmitted through the light in a room for laptops, smartphones, and tablets. In the proposed system, LED light emits visible light with location data to the visible light receiver and the receiver receives the data.

The embedded system calculates the optimal path to a destination and speaks to the visually impaired through a headphone or speaker.

## 2. Li fi technology

The idea of Li-Fi was first given by Harald Haas from University of Edinburgh, UK, in his TED Global talks on Visible Light Communications. According to him Visible Light Communication is very simple, if the LED is on, the transmitted data is digit 1, if it is off, the transmitted data is digit 0. The LEDs can be switched on and off very quickly, which gives better opportunity for transmitting data. So the requirement in LEDs and a controller that code data into those LEDs (**Figure 1**) [1].

Depending on the data to be encoded or transmitted, the LED flicker rate is varied. Enhancements to be made in this method are like using combinations of red, green and blue LEDs or using parallel data transmission LED array to change the light's frequency with each frequency encoding a different data channel. By the above advancement it is possible to achieve a theoretical value of speed up to 10 Gbps, i.e. downloading a 1Gb file in just 30 seconds irrespective of the file format. It can be used in places such as hospitals, traffic signals and in modern medical instruments. Li-Fi can also be used underwater where 6 Wi-Fi [2].

In transportation applications, Vehicle-to-vehicle (V2V) communications are deliberated to play an important role in in the next decade to improve road safety and road capacity [1]. In recent years, V2V has been implemented using the mainstream technology called as Dedicated Short Range Communications (DSRC), a 5.9 GHz radio frequency (RF) technology. However, the conventional RF communications based V2V communications often agonized from low packet reception rate and long delay in high vehicle density scenarios, due to interference made by the huge number of nodes in the same network [3]. On the other hand, it is very difficult to visually identify the location of the transmitter sending a message is frequently difficult, as RF based transmissions are usually Omni directional and latest technologies have inadequate accuracy to support this. For vehicle localization, the most common technology used is GPS. However, the positioning error occurred in GPS devices is often more than 10 m creates it hard to find the transmitting vehicles when they are in close locations [4].

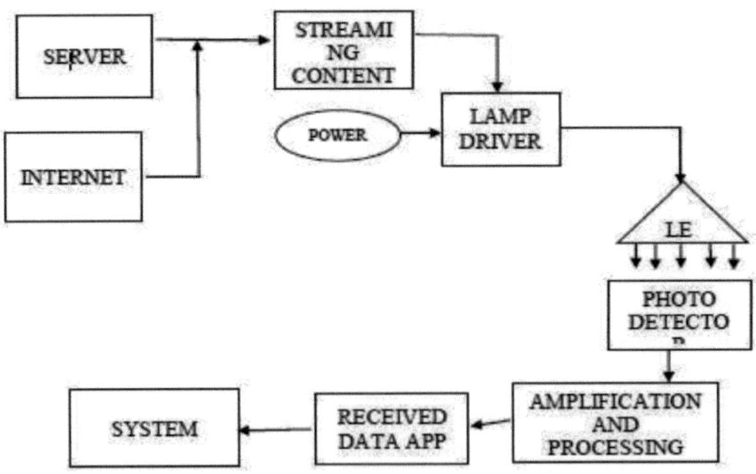


**Figure 1.**
*Block diagram of Li-Fi.*

## 3. Li fi –transmitter and receiver

The operational procedure of Li-Fi is very simple, it transmits binary 1, if the LED is in on condition, if the LED is off, it transmits a binary 0. The switching of LEDs between on and off positions can be implemented very quickly, which gives wonderful chances for data transmission. A controller which is used to program the LEDs and some LEDs are mainly needed. To implement the Li-Fi technology, it is required to just vary the LED's flicker rate based on the data we need to encode. All the data on the internet will be streamed on one end to a lamp driver when the LED is switched on the controller which converts the digital data in form of light. At the receiver end, the received signal is converted back to original signal with the help of a light sensitive device (photo detector). This method of wireless transmission of data using rapid pulses of light is technically denoted as Visible Light Communication (**Figures 2** and **3**) [5].

**Figure 2.**
*Top view of Li-Fi transmitter and receiver.*

**Figure 3.**
*Bottom view of Li-Fi transmitter and receiver.*

Light based Wi-Fi is called as Li-Fi. It uses light signal instead of radio waves for data transmission. Li-Fi uses transceiver-fitted LED lamps instead of Wi-Fi modems. Transceiver-fitted LED lamps can be used to light a room as well as transmit and receive information [5]. Power supply is used to provide power to the transmitter and receiver section. Data stored in the controller is transmitted through a Li-Fi LCD display that is used to check whether the data is transmitted and received. Microcontroller is used to control both transmitter and receiver sections. A controller used in both transmitter side and receiver side. Using Proteus software tool program written to microcontroller. Li-Fi is a term used to describe the visible light communication technology applied to high speed wireless communication (**Figures 4** and **5**).

**Figure 4.**
*Working module of Li-Fi Tx and Rx via LED.*

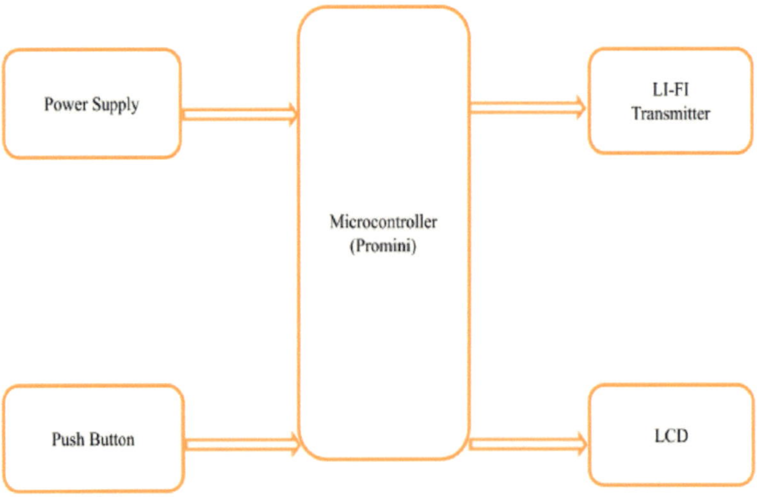


**Figure 5.**
*Li-Fi transmitter.*

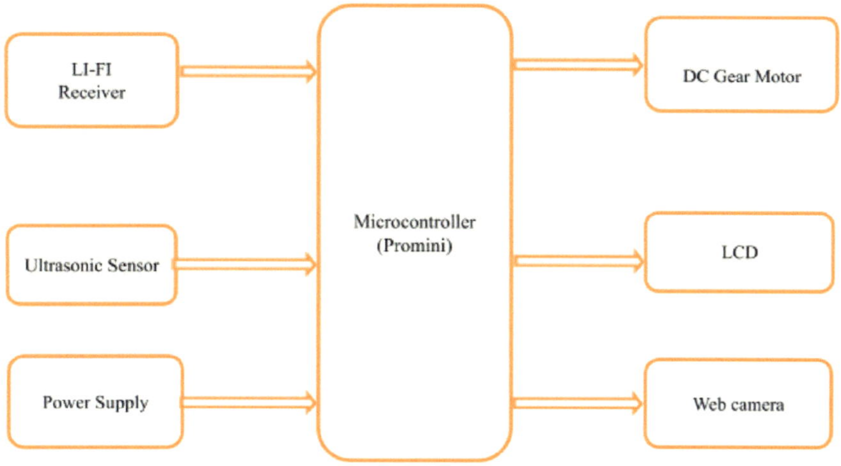


**Figure 6.**
*Li-Fi receiver.*

Receiver consists of Photo diode which receives the signal transmitted by transmitter. Transmit the received signal to the controller. FN-M16P module is a serial MP3 module which includes a perfect integrated WMV and MP3 decoder chip. Further, it includes micro SD card drivers and supports FAT32 and FAT16 file systems. It is able to play back specified sound files. Speaker or earphone is used to give audio signal to impaired person(**Figure 6**) [6].

## 4. Arduino pro mini – Microcontroller

Arduino Pro Mini is used in the system for application module. It does not have inherent programmer since it is an application module. Various connectors and USB port are useless and they are removed from the module because the module is connected with the application programmer. There are two versions of Arduino Pro Mini. They are classified based on the working voltage of controller. The voltage levels are +3.3 V and + 5 V. Based on the application, the designer can choose the appropriate board.

In recent years, Arduino boards are commonly used because its operation and architecture is easy to understand. Also the original module schematics and required software modules related to Arduino is available as an open source platform. Based on the requirement designer can customize the system using this open source platform. Different types of Arduino boards are available on the market for designing a system [7]. They are accessible with various packages and features. Appropriate boards can be chosen depending on the requirement.

The few reasons why pro mini is selected over other are listed below:

Case1: Permanent installation of the system is used. Only the board is required to be programmed once in permanent applications. In such systems, the features like a USB programmer, I/O connectors and other supporting hardware is not required. The pro mini is explicitly designed for those systems which uses permanent installation. It has some basic hardware modules that is just enough for those applications.

Case2: This pro mini board is one of the smallest boards of Arduino. Because of its small size it can be used in mobile applications.

Case3: The cost of the board is significantly lesser due to its basic hardware structure.

Case4: It can able to store most application programs in its 32Kbytes memory.

The operation of Pro Mini is similar to any other development board. All you have to do is to write the controller program and select the appropriate interfacing modules to get system running. The detailed step for the programming of pro mini is given below.

1. To program the pro mini, you cannot connect directly to PC because it does not have an inbuilt programmer. Select either SPI or UART programmer. UART programmer can be preferred.

2. Download ARDUINO IDE software and install the same in your computer.

3. List the roles to be accomplished by pro mini.

4. Write the functions as 'C' language programs in IDE.

5. Establish a communication between IDE and pro mini by connecting the programmer.

6. Burn the program and download to pro mini through IDE.

Remove the programmer module. Provide the power and connect the necessary peripherals. The desired output is obtained after resetting the control and executing the program.

## 5. Hardware components

### 5.1 Gear motor

A DC motor can be defined as the rotary electric motor that converts direct current electrical energy into mechanical energy. The forces produced by magnetic fields mainly define its types. In order to change the direction of current, all types of DC motors have some internal mechanism, either electromechanical or electronic periodically.

In robotics, DC motors are most commonly used as it is available in a large variety of shapes and size with permanent magnet iron core, permanent magnet

**Figure 7.**
*Gear motor.*

ironless rotor, permanent magnet brushless, wound field series connected, shunt connected, compound connected, variable reluctance stepper, permanent magnet stepper, and hybrid stepper motors (**Figure 7**).

The characterization of DC motor is based on brush type and brushless. It refers to the design of commutation used in motor, which converts direct current from the batteries into alternating current which are required to generate the action of motor. The brush type DC motors performs mechanically with brushes with the commutator segments at the ends of the rotating rotor physically slides against the stationary brushes that are connected to the terminals of the motor. In brushless motors, DC is converted to AC electronically with the position of sensors and microcontroller in the rotor; hence it does not require brushes [8]. A conventional DC motor is formed by an arrangement of coils and magnets that creates motion from electrical power [9].

### 5.2 Push button switches

Push button switches are small and sealed and the circuit gets completed by pressing it. If it is ON, a small metal spring inside makes contact with the two wires, passing the electricity to flow through the circuit. If it is OFF, means the spring retracts and the circuit is open, thus not allowing the current to flow through it. These buttons are made of hard material such as plastic or metal.

A simple Push Button switch is a type in which the switch consists of electric mechanism or air switch mechanism to turn ON or OFF. Based on the model of switch, the operations can be given as momentary or latching action function. The button is usually made of a strong durable material such as metal or plastic [9]. A push to make switch allows electricity to flow between its two contacts when held in. When the button is released, the circuit is broken. This type of switch is also known as a Normally Open (NO) Switch.

### 5.3 Ultrasonic sensors

An ultrasonic sensor is an instrument that uses ultrasonic sound waves to measure the distance of an object. It uses a transducer to send and receive the ultrasonic pulses that relay back information about an object's proximity. The distinct echo patterns are produced by the high-frequency sound waves reflected from boundaries.

The HC-SR04 Ultrasonic (US) sensor is a 4 pin module (Vcc, Trigger, Echo and Ground) respectively. This system has two features in the front which forms the Ultrasonic transmitter and Receiver. The sensor works based on the simple formula i.e. Distance = Speed × Time. The Ultrasonic transmitter transmits an ultrasonic wave, this wave travels in air and when it gets objected by any material it gets reflected back toward the sensor this reflected wave is observed by the Ultrasonic receiver module. HC-SR04 distance sensor is commonly used with both microcontroller and microprocessor platforms like Arduino, ARM, PIC, Raspberry Pie etc. The following guide is universally since it has to be followed irrespective of the type of computational device used. Power the Sensor using a regulated +5 V through the Vcc ad Ground pins of the sensor. The current consumed by the sensor is less than 15 mA and hence can be directly.

It is powered by 5 V pins and the Trigger and the Echo pins are both I/O pins. It can be connected to I/O pins of the microcontroller and to measure the trigger pin has to be made high for 10uS before turning off. This function instantly triggers an ultrasonic wave at a frequency of 40 Hz from the transmitter. At the same time, the receiver had to wait for the signal to return [10]. The wave is returned after it is reflected by any object; here the echo pin will be high for a particular amount of time, which is equal to the time taken for the signal to return back to the sensor. Hence, the amount of time during the high position of Echo is measured by the MCU/MPU. It gives the measured value for the time taken by the wave to return back. Using this values the distance is measured as explained above.

The ultrasonic Sensors are efficiently used in the non-contact detection of Level, Position, and Distance. These non-contact sensors are also called as proximity sensors. It is also independent to light, smoke, dust and color. The material absorbs the ultrasonic sound wave and does not reflect sound except wool.

### *5.3.1 Ultrasonic distance measuring*

Distance measurement is estimated with respect to the measurement of time-of-flight. The accurate distance is calculated with the time taken from sending the ultrasonic sound and receiving the reflected sound signal. In application such as, height monitoring, bin level measurement and proximity zone detection various ultrasonic sensor like MB7360 HRXL-Max Sonar-WR are used to obtain precise measurement.

- Ex. Distance measurement in a garage parking application uses the sensing method when a vehicle is pulled completely into a garage.

- An ultrasonic sensor, MB7360 occupies as bin level sensor to identify the materials in bins.

### *5.3.2 Ultrasonic obstacle detection*

- The design of Unmanned Aerial vehicles and robots use ultrasonic sensor and proximity sensors for obstacle detection.

- Ultrasonic sensors are more precise for close range detection up to ten meters and provides multiple range measurement per second.

## 5.4 Webcam

A webcam is used to capture live video or image through computer interfaced network. Webcams are more compatible with user's monitor and hardware interface. Webcams finds popular application in video chat session with live audio and video streaming. The device such as Apple's iSight camera, iMacs webcams provides clear picture for video chat sessions. The software supported with Webcam records the video and also it can be live streamed through Internet. The video streaming over the Internet needs a wide bandwidth and it is compressed for live streaming. The maximum resolution of a webcam is lower compared with handheld video cameras and it suits for video chat session. Webcam are used to capture the view of camera over its web page. The evolution of webcam technology uses advanced webcams with the improvement from 1080 pixel to 2160 pixel for the better image quality and widely used in several industries including marketing, security, traffic management and healthcare systems.

### *5.4.1 Working of the webcam*

Webcam works as a conventional digital camera to interact with the web pages and other internet pages. The camera technology is framed to capture the images through a tiny grid of light-detectors, known as charge-coupled devices (CCD). The output of CCD is the digital format so that computers can access this data. Webcams do not provide the option of image storage and it transmits the data immediately to the host device through the USB cable. The advancement in technology leads to integration of microphone with camera. Webcam function in two ways as capture the image or video and to transfer it to the predestined device [9]. Similar as digital camera, the webcam has come with the appropriate software to interact with the host device. The Software provides the flexibility to edit the images and to record the videos for particular duration. This software collects the digital data of image from the camera at certain intervals of time. The frame rate decides the number of pictures or video streaming displayed on the computer or other display systems. The image frame is converted into a JPEG file and finally sends it to the web server using the file transfer protocol (FTP). To utilize the webcam in transferring the data, certain configuration steps to be followed in uploading the images and videos [10].

Modern day desktop and laptops are provided with built in webcam of small in size. Due to this compactible size, the provision for multi-piece lens is not possible and this results in the reduced image quality. To overcome the drawback in image quality, external webcams are chosen. Adjusting the camera position is difficult as it is fixed with the device [11]. The integrated webcam and external webcams are varied with respect to cost, focal length, stereo quality sound, light sensitivity and certification.

#### *5.4.2 Image sensor*

Image sensors are classified based on its structure type as CMOS or CCD. In CMOS image sensor uses a solid-state image sensor chip which is made up of light sensitive elements, micro lenses, and micro electrical components. In CCD the pixel exposure occurs at the same time in the technology of global shutters. The webcams are capable of providing VGA-resolution with the frame rate of 30 frames per second. Modern day devices are developed to produce video in multi-megapixel resolutions such as the PlayStation Eye, which can produce 320 × 240 video at 120 frames per second. Nintendo Wii Remote uses an image sensor with a resolution of 1024 × 768 pixels [10].

#### *5.4.3 Webcam Interface*

Typical interfaces such as Ethernet and IEEE 802.11 (denominated as IP camera) are used for webcam connectivity for desktop or Wi-Fi devices. To communicate with high quality video through single channel or with two or more channels, the interface such as composite video or S-Video interface is used. The video streaming functionality to USB enabled device is achieved through USB video class (UVC) specification to interface effectually with host machines.

#### *5.4.4 Webcam features*

##### *5.4.4.1 Frame rate*

The webcam with good image quality is obtained with the frame rate of 30 frames per second (fps). The frame rate less than 30fps leads to blurred in image quality. It is better to choose webcam which that supports 60 fps recording to obtain decent image quality. The increase in fbs rate occupies the modern day monitors to capture live image and the fps rate decides the speed at which the image moves on the screen. Webcam image is transferred with a range of 15 frames per second to an ideal rate of 30 fps and this provides better video streaming with respect to the speed of the internet.

#### *5.4.4.2 Resolution*

The resolution of webcam has improved to 720 × 1080 pixel to meet high-definition capabilities. The higher end webcams are also available as 4 k range, which can be used for specific applications and it is supported with HD-capable monitor to view the true high definition. Webcams with 1080p are compatible with all interfacing devices to perform real time video streaming.

#### *5.4.4.3 Autofocus*

Autofocus is an option provided in webcam, which is used to focus the object automatically from camera position. Autofocus mode creates a delay for image capture as the camera takes the time to focus the object. In order to speed up the video streaming then the auto focus mode is turned to off condition.

#### *5.4.4.4 Microphone*

High-quality recording for episodes or other higher-tech films requires an upgrade. For those situations, invest in an external microphone.

#### *5.4.4.5 Megapixels*

It decides the quality of the picture or image. Most of the cameras provide reasonable quality images. It is good if we use 320*240 or 640*480 pixels. For better quality webcam should have 1280*720 resolutions.

#### *5.4.4.6 Lens*

Webcams are designed with glass lens or plastic lens based on the pixel range required and the cost convenience. For professional video presentation through Skype and other video chat software can use webcams with plastic lens, as it is more adequate for decent image quality [12].

### *5.4.5 Webcam resolution test*

The Webcam Resolution Test is used to estimate the resolution of webcam through internet connectivity. The test results will show the clear list of resolutions supported by the camera and provides the information for attaining the ways of maximum resolution, minimum resolution and default resolution. The analysis provides the support to estimate each supported resolution for the image taken and each image quality can be compared. The webcam resolution test initially checks as default mode for possible resolution standards the major drawback occurs if the supported resolution is missed out.

The pixel value of captured image depends on the height and width of captured image. The image quality is specified as megapixel as one million pixel equals to one megapixel and abbreviated as Mpx or MPixel. The image quality of camera is determined based on the megapixel range as the manufacturer indicates the maximum supported value [13]. An webcam with a resolution of 1920 × 1080 can capture up to 2073600 pixels or 2.0736 megapixels, rounded off as 2MP. The resolution of camera is not the only parameter for better image quality as higher the resolution provides more accurate details and good sharpness. The image taken with a higher resolution camera finds more clarity, when printing in large formats or viewing on big screens.

## 6. Result and discussion

Light Fidelity is a high speed wireless communication through light-emitting diodes leaf and Wi-Fi are quite similar as both transmit data electromagnetically however Wi-Fi gives radio waves, while Li-Fi transforms visible light waves. Visible light communication (VLC) accommodates a photo detector to receive light signals

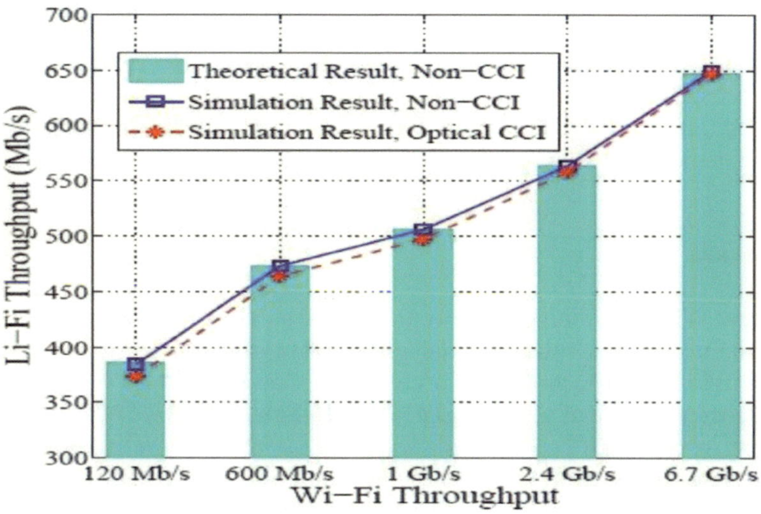


**Figure 8.**
*Simulation results of Wi-Fi vs. Li-Fi.*

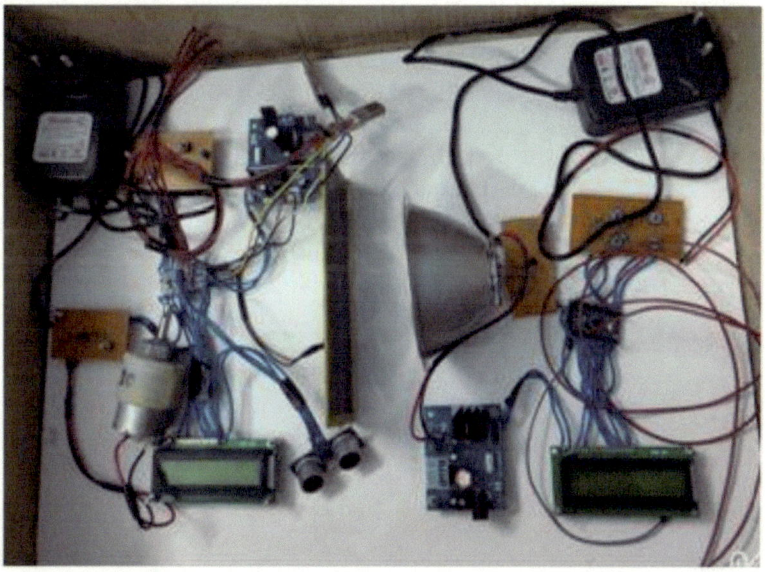

**Figure 9.**
*Hardware model of proposed system.*

and uses a signal processing elements to convert the data into stream-able content. A semiconductor light source is operated at extremely high speed to perform visible light communication. The light source is dimmed below human visibility to carry the data [14]. For short distance communication can be preferred with this and it is possible in electromagnetic sensitive areas. The proposed model uses a leaf architecture with two modules, in which one act as leaf transmitter and the other as leaf receiver (**Figure 8**).

At transmitting sides we have buttons, when we press the right button the microcontroller will send data through the leaf and the module at the receiver receives the data by the leaf and sends it to the microcontroller. At the transmitter side the data transmitted through UART communication and receiver side we have a solar band receiver [11]. So that both the data are displayed on both by LCD modules. According to the instructions they have been programmed to display right, left, brake instructions by the vehicles. The proposed system hasa face recognition safety authorization system. The ultrasonic sensor detects the vehicles which are less than threshold distance and sends an alert message by displaying brake in the LCD display. The relay will apply the brake on the second vehicle automatically when the data is received (**Figure 9**).

## 7. Conclusion

Similar as Wi-Fi technology, Li-Fi can be used as a bi-directional wireless communication method. Li-Fi uses visible light for data transmission as the RF communication is used for Wi-Fi and cellular networks. Compared with Wi-Fi technology for data transfer the use of visible light makes Li-Fi technology 100 times faster with minimum cost. Li-Fi communication can be performed in low power mode and no external power source is required as it operates with the glowing LED light. Li-Fi finds major advantage in electromagnetic sensitive areas, such as hospitals, aircrafts, and nuclear power plants, as it does not cause any electromagnetic interference. The technical advantages of Li-Fi drive its market in various application sector such as retail, aerospace, defense and indoor networking. These technical advancement over LI-Fi communication finds a great demand in the market within next few years.

We have presented a Visible Light Communication (VLC) system designed with Li-Fi transmitter and receiver introduced its characteristics and capabilities for precise communication. The VLC system in traffic signal are modeled to communicate with the LED lights of the cars and the traffic information can be shared with the vehicle which further reduces the accident [15]. Li-fi is more suitable for high density coverage with in a confined region and the technology yields a high speed communication of 10 Gbps. Li-Fi technology will advance to replace Wi-Fi communication for high speed data transfer application in near future.

## Author details

P.M. Benson Mansingh*, G. Sekar and T. Joby Titus
Sri Ramakrishna Institute of Technology, Coimbatore, India

*Address all correspondence to: benson.ece@srit.org

## References

[1] Y. Goto, I. Takai, T. Yamazato, H. Okada, T. Fujii, S. Kawahito, S. Arai, T.Yendo, and K. Kamakura, "A new automotive VLC system using OpticalCommunication image sensor," IEEE Photonics Journal, vol. 8, no. 3, pp. 1-17,2016.

[2] K. Siddiqi, A. D. Raza, and S. S. Muhammad, "Visible light communication forV2V intelligent transport system," in 2016International Conference onBroadband Communications for Next Generation Networks and MultimediaApplications (CoBCom). IEEE, 2016, pp. 1-4.

[3] S.-H. Yu, O. Shih, H.-M. Tsai, and R. Roberts, "Smart automotive lighting forvehicle safety," IEEECommunication sMagazine,vol.51,no.12,pp.50– 59,2013.

[4] D.R.Kim,S.H.Yang,H.S.Kim,Y.H.Son, andS.K.Han,"Outdoor Visible light communication for inter-vehicle communication using controller area network,"in 2012 FourthInternational Conference on Communications and Electronics(ICCE), Aug 2012, pp.31-34.

[5] A.-L.Chen,H.-P.Wu,Y.-L. Wei,andH.-M.Tsai,"Time Variation Vehicleto-vehicle visible light communication channels,"in Proc. IEEE VehicularNetworking Conference (VNC), Dec 2016, pp.1-8.

[6] P. Luo, Z. Ghassemlooy, H. L. Minh, E. Bentley, A. Burton, and X. Tang,"Performance analysis of a car-to-car visible lightcommunication system," Applied Optics, vol. 54, no. 7, pp. 1696-1706, Mar2015.

[7] P. M. B. Mansingh, T. J. Titus and V. S. S. Devi, "A Secured Biometric Voting System Using RFID Linked with the Aadhar Database," *2020 6th international conference on advanced computing and communication systems (ICACCS)*, Coimbatore, India, 2020, pp. 1116-1119, doi: 10.1109/ICACCS48705.2020.9074281.

[8] Mansingh, P.M.B., Yuvaraju, M. Improved data transmission using Li-Fi technology for home automation application. J Ambient Intell Human Comput 12(5), 5581-5588 (2021).

[9] "Li-Fi Based A New Home Automation System", P.M Benson Mansingh, M. Nithya, M.Krithika, Department of Computer Science, Anna University, IJRASET Volume 4 Issue II, February 2016IC Value: 13.98 ISSN: 2321-9653.

[10] P M Benson Mansingh *et al* 2020 *IOP Conf. Ser.: Mater. Sci. Eng.* **932** 012091,Ber Analysis Of Channel Coding Techniques For 5G Networks, IOP Conference Series: Materials Science and Engineering, Volume 932, 1st International Conference on Science, Engineering and Technology (ICSET) 2020 27 February 2020, Penang, Malaysia.

[11] S. M. Motaghiannezam et al., "Single chip 52 Gb/s PAM4 transmission through −58 and +10ps/nm chromatic dispersion using directly modulated laser," in Proc. Conf. Opt. FiberCommun.,2016, Paper Th2A.59.

[12] P. Zhang and Q. Li, "Performance of UWB impulse radio with planar monopoles over on-human-body Propagation Channel for wireless body area networks," IEEE Transaction onAntennas and Propagation, Vol. 55, No. 10, October 2007, pp. 2907-2914.

[13] H. H. Lu et al., "45 Gb/s PAM4 transmission based on VCSEL with light injection andoptoelectronic feedback techniques," Opt. Lett., vol. 41, no. 21, pp. 5023-5026, 2016.

[14] S. Lange et al., "Low switching voltage Mach–Zehnder modulator

monolithically integrated withDFB laser for data transmission up to 107.4 Gb/s," J. Lightw. Technol., vol. 34, no. 2, pp. 401-406,Jan. 2016.

[15] S. Bates, D. Elliot, and R. Swamy, "Termination sequence generation circuits for low-densityparity-check convolutional codes," submitted to IEEE Trans. Circuits and Systems I, March 2005.

Chapter 4

# The Role of Ubiquitous Computing in the Transformation of the Healthcare System

*Ana Lídia Corrêa da Silva Moreira*

## Abstract

Currently, keeping the financial balance in health services has been a challenge for governments and private companies. Factors such as population aging, medical inflation and the lack of adherence to treatments raise costs and stimulate debates around the world on how to maintain systems' budget viability. During the last decades, we could observe the phenomenon of computation needing less hardware, becoming more portable and migrating to our offices, pockets, clothing, and finally, to our body, thus allowing us to envision a range of solutions for these problems. The only way for society to offer such quality services and overcome these challenges is through the solutions that the convergence of new technologies brings to us now. The applications of ubiquitous computing have the potential to immensely benefit patients, managers and the society that finances the health system.

**Keywords:** Ubiquitous Computing, Healthcare, Population Aging, Adherence

## 1. Introduction

> *"The most profound technologies are those that disappear. They weave themselves into the fabric of everyday life until they are indistinguishable from it [1]".*

Our time is marked by cheap sensors and fast and powerful processors. These elements have the power to cause profound changes in our society and unleash great opportunities, which lead us to the Ubiquitous Computing field. Mark Weiser defined Ubiquitous Computing as computation (not necessarily computers) everywhere and everything. Although, at the moment, Weiser considers that computers were still "in a world of its own [1]", he already envisioned a new threshold, when the technological advances of computing would overcome the computer to install itself in the most common objects of the daily life [1, 2]. In this sense, it is noteworthy to mention the Internet of Things (IoT) technology as an important element of the Ubiquitous Computing reality.

During the last decades, we could observe the phenomenon of computation needing less hardware, becoming more portable and migrating to our offices, pockets, clothing, and finally, to our body [3]. This spreading process brings in-depth transformations to all human activities and healthcare, and all its dimensions are being strongly impacted. Throughout the world, the healthcare systems, including governments and private companies, are facing challenges to keep the financial balance. Regarding this issue, there are two causes I would like to highlight: Population aging and the lack of adherence to health treatments [4].

### 1.1 Population aging

We can easily observe a fast change in the demographic profile of the world population. The drop of mortality added to the increase in birth rates in the 40's and 50's resulted in a rise of people aged 60 and older [5]. The phenomenon of the population aging can be observed in all the countries, even in regions with low- and middle-income [5]. These rapid and profound changes in the age pyramid also means transformations in various aspects of society, including family dynamics, social security and healthcare demands. Therefore, it is possible to affirm that society is experiencing "a longevity revolution" [5].

The longevity revolution is a gift to our generation and future ones. But, for enjoying all the benefits of this gift, is essential that society is prepared to offer to individuals a healthy and active aging process, allowing people to keep independence and quality of life as long as possible. Thereby, in order to elaborate appropriate plans to achieve such an important and challenging goal, it is crucial to understand the size and speed of the aging process in society. **Figure 1** highlights that the percentage of persons aged 65 years and older increased from 6 per cent in 1990 to 9 per cent in 2019. According to United Nations projections, this estimate is expected to reach 16 per cent by 2050. Additionally, life expectancy at birth reached 72.3 years [6].

Another interesting data from the United Nations report is the number of persons aged 80 years or older. In **Figure 2**, we can observe that this age group raised from 54 million in 1990 to 143 million in 2019. Moreover, the estimates indicate they will be 426 million by 2050 [6]. Therefore, we can expect the costs of healthcare to increase, owing to the strong correlation between population aging and healthcare expenditure, being this impact on health finances largely propelled by the prevalence of chronic conditions in elderly population [7, 8].

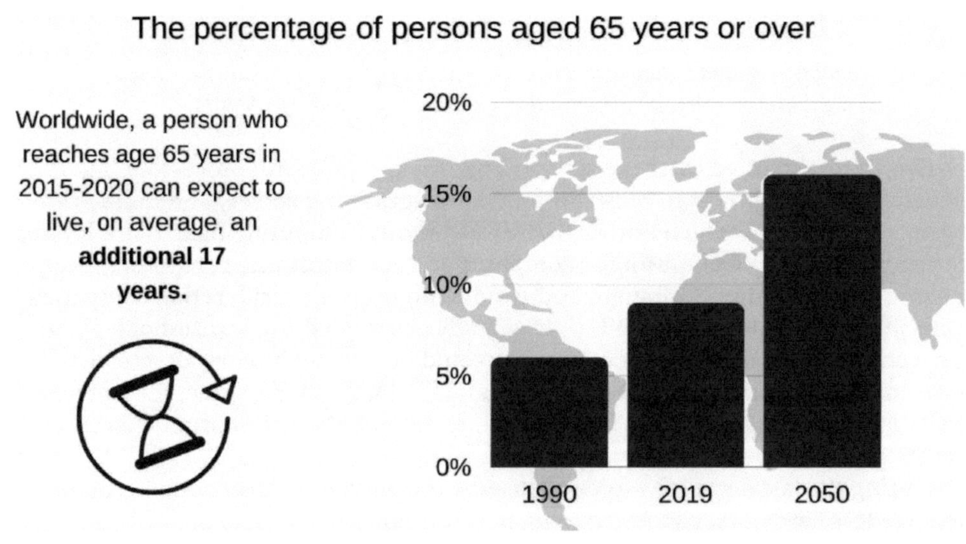


**Figure 1.**
*Percentage of persons aged 65 years or over from 1990 to 2050.*

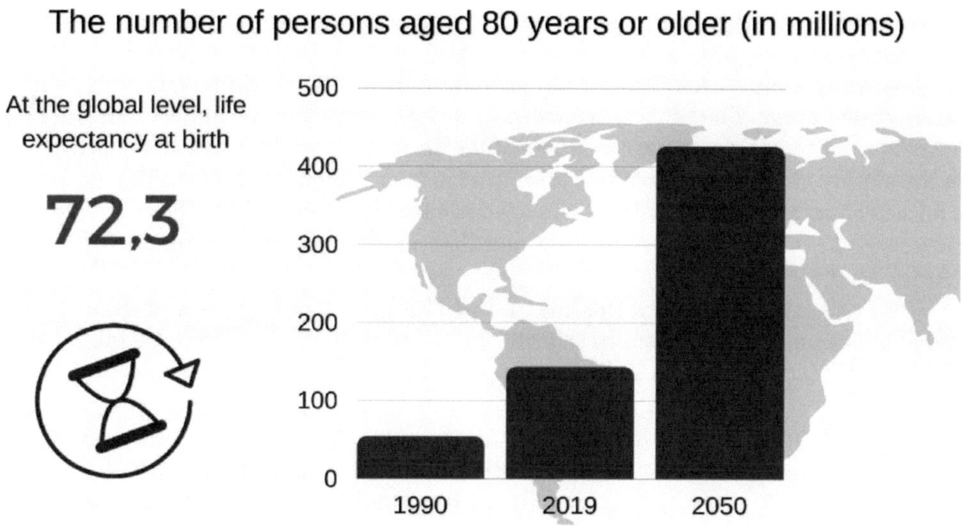


**Figure 2.**
*Number (in millions) of persons aged 80 years or over from 1990 to 2050.*

### 1.2 Adherence to health treatments

To start the discussion, it is pivotal to understand the meaning of adherence to health care practices. The term suffered an evolution, abandoning the simple idea of adherence to medication or even "the extent to which the patient follows medical instructions" [9] to include a wider range of individual behaviors. According to World Health Organization (WHO), the active search for treatment, immunization, self-management of risk habits and several other actions can be included in the list of indicators of adherence to healthy habits [9].

In fact, adherence is a very complex and multifaceted term, involving a diverse range of actions, behaviors and habits of the individuals. Thus, the task of finding an appropriate and complete definition for adherence already proves to be challenging, evidencing the possible difficulties to choose the most suitable indicators to measure it.

Despite these concerns and the risk of incurring in results with a certain degree of inaccuracy, it is important to understand the adherence and how to improve the levels of commitment and engagement of the patients, since a low adherence has negative impacts on the effectiveness and cost of health treatments [10].

Most of the attempts to measure adherence rely on patients' reports, which tend to be incomplete and subjective, especially when the patient is non-adherent [10]. However, the recent technological advances give us the unique opportunity to develop extremely precise tools to evaluate how much the behaviors of the individuals draw near the healthcare professional recommendations and, finally, to elaborate strategies based on reality.

Thus, it is possible to realize the association between the transformation of healthcare demands and the possibilities of ubiquitous computing in promoting the transformation of healthcare paradigms. The emerging technologies that compose this new ongoing reality, known as ubiquitous computing, have the potential to cause great revolutions not only for patients, but mainly for managers and for the society that finances the health system.

## 2. Internet of things

Regarding Ubiquitous Computing paradigm, it is worthwhile to talk about the Internet of Things. The term was first used in 1999 during a presentation performed by Kevin Ashton about the application of Radio-Frequency Identification (RFID) in the supply chain. In this context, the Internet of Things was born in the industrial sector or, at least, strongly stimulated to meet the demands of the industry. However, the technology is spreading rapidly and affecting several sectors of human activity [11].

After this start, gradually, the Internet of Things has undergone a profound transformation and expanded to refer to the concept of object networks, including everyday objects, "that are readable, recognizable, locatable, addressable, and/or controllable via the Internet [12]". The communication technologies used to connect these objects are not relevant to our definition. Further, it is crucial to enlarge the concept beyond electronic devices and other products with high added technology, such as vehicles or home appliances. The concept embraces less obvious things, as well as food, clothing and soil [12].

**Figure 3** reveals that the expectation is that the number of connected objects continues to increase. Currently, there are already more than 12 billion of IoT connected objects worldwide, and this number should exceed 25 billion by 2025, hence representing an increase of 100%. Another interesting fact is that there is an inversion, at the moment, since there are more connected objects with consumers than with industry. The consumer's objects are expected to grow 63%, whereas the objects of the industry are expected to grow 145% over the next five years. So, we can notice a vigorous movement of the productive sector to incorporate this new technology [13].

Considering the IoT spending, it is expected that the amount will reach USD 1 trillion by 2022. The current values are near USD 750 billion and it is foreseen an improvement of 34% in the next two years, which proves how thriving this market is [14].

### 2.1 Internet of medical things

As previously mentioned, IoT is causing a revolution in society. Obviously, the healthcare sector is part of it and is reaping great benefits from the implementation

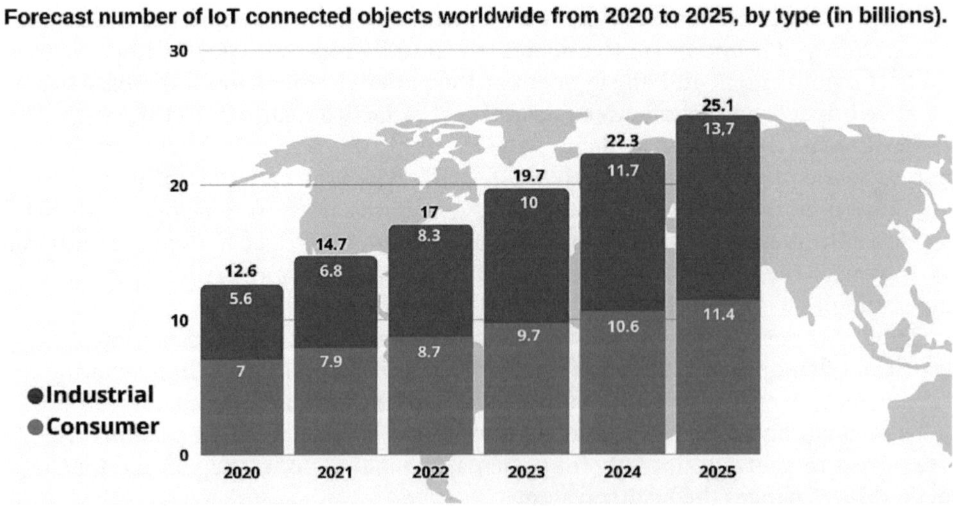


**Figure 3.**
*Number (in billion) of IoT connected objects worldwide from 2020 to 2025.*

of this technology. Interestingly, IoT originated the term Internet of Medical Things (IoMT), which describes connected medical devices. It is estimated that the overall IoMT market will reach USD 158 billion by 2022 [15]. In addition, people are taking the initiative to use smartphones, wearables or other devices to support the monitoring and management of healthcare. **Figure 4** shows that most users are interested in information related to maintenance of good health and exercise level, for example, thus indicating how powerful IoMT can be in preventive medicine [16].

These data highlight the role of IoMT in the transformation of the healthcare system to an individualized and patient-centered system (eHealth), in opposition to the current one based on hospitals and mass services [17, 18]. We are observing the growing number of smart and connected medical devices "able to generate, collect, analyze or transmit health data or images and connect to health care provider networks, transmitting data to either a cloud repository or internal servers [15]".

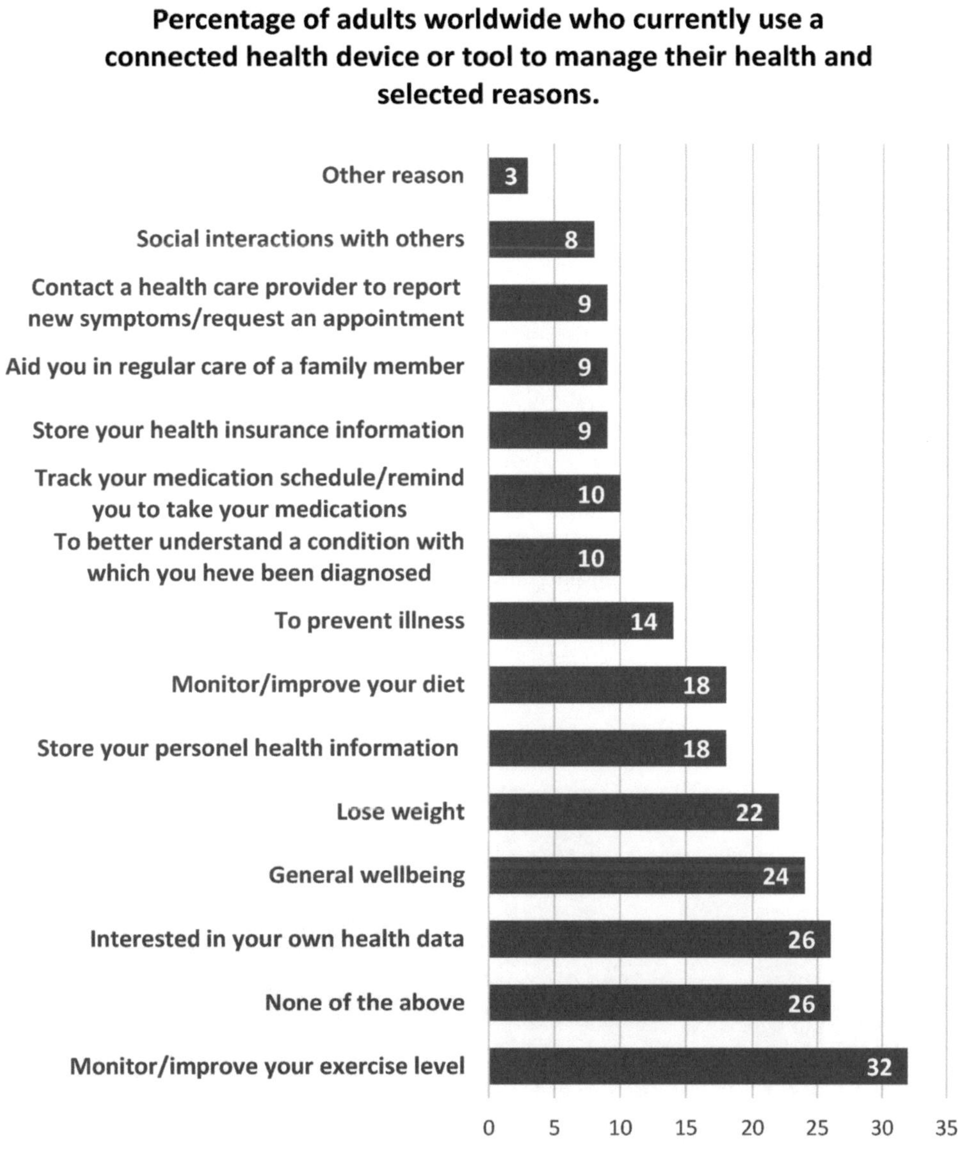


**Figure 4.**
*Percentage of adults worldwide who currently use a connected healthy device or tool to manage their health and selected reasons.*

Accordingly, these devices represent a new territory with the possibility to create a fresh and innovative paradigm of connected system and data-drive healthcare. The upcoming changes have the potential to drop the costs and increase access to health care.

It is worthwhile mentioning that the COVID-19 crisis accelerated the adoption of IoMT due to increased pressure on the health system and high risk of contamination in hospitals. Remarkably, IoT improves efficiency in all the health systems through the integration of the network and the usage of all sorts of data in real time to better manage available resources, which is highly important at this moment of health emergency [17, 19].

Furthermore, numerous solutions able to reduce the circulation of patients in the hospitals are gaining attention, and there has been a rapid growth of technologies related to telemedicine and patient remote monitoring [17, 19]. Two main lines of application are here observed. First, the monitoring of COVID-19 patients to follow up the evolution of the disease and reduce the need for circulation of an infected individual. Also, there are examples of IoMT usage for monitoring of patients with chronic disease, because they are a risk group for COVID-19, but it is not possible to simply waive a regular medical supervision [17, 19].

## 3. Ubiquitous computing and healthcare: a bibliometric analysis

A useful tool to better understand the extension and importance of Ubiquitous Computing in healthcare is the assessment of academic publications on this topic. The efforts undertaken by Universities and Research Centers to advance the technology and develop new possibilities of application demonstrate the expectations that society deposits on it, and also exhibit the potential that the field has to advance propelled by investments in research and development.

A search was conducted in the National Center for Biotechnology Information, U.S. National Library of Medicine database using the terms [(ubiquitous computing

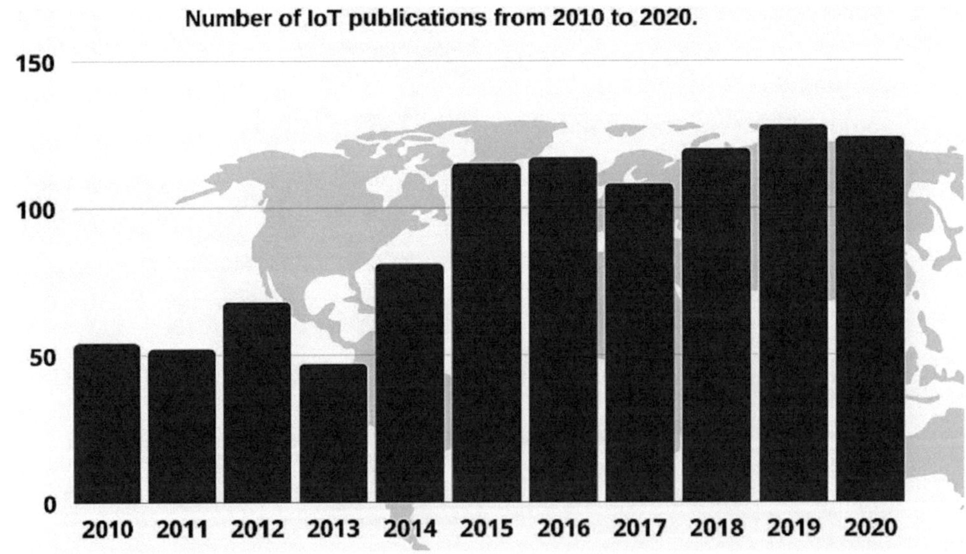


**Figure 5.**
*Number of publications retrieved from the National Center for biotechnology information, U.S. National Library of Medicine database using the terms [(ubiquitous computing AND (health OR healthcare)]. Publications from 2010 to 2020 were considered.*

AND (health OR healthcare)]. Publications from 2010 to 2020 were retrieved. The search returned 913 results. **Figure 5** depicts the distribution of these publications in the timeline. Interestingly, an important growth between 2013 and 2015 is evident. Besides, there is a constant improvement in the number of publications every year. In turn, the amount of publications demonstrates that there is space to better explore the subject.

## 4. Conclusions

Based on the information exposed, we can conclude that the healthcare systems are under an important pressure caused by the population aging. Even countries with low- and middle-income are experiencing the rise of life expectancy. Obviously, this indicates advances in many social aspects, but also shows that society must be prepared to offer healthcare and life quality for this growing population group. It is crucial to develop policies to help people stay healthy and living independently as long as possible.

However, to achieve this goal, the challenges related to the adherence to treatments must be faced. Despite the improvement of healthcare access, there is still a lack of adherence to healthcare treatments and healthy habits, including exercise, for example. Indeed, the low adherence impacts negatively the effectiveness and cost of health treatments.

At the same time, despite these important challenges for healthcare providers, we also have a very unique opportunity to come up with the evolution and convergence of technology. The concept of ubiquitous computing is deeply involved with the IoT technology and its applications. After all, the vision of computation everywhere and everything is becoming reality through this growing network of smart objects.

These new technologies are proven to be a new and prosperous economic sector, since the number of connected devices is rapidly rising, as soon as the respective spending. Besides that, for the first time in medical history, it is possible to understand all aspects of the elements able to interfere in the population's health, from the functioning of the hospitals to individual behaviors in daily life. Finally, access to such information will transform the way we think and offer healthcare services.

## Author details

Ana Lídia Corrêa da Silva Moreira
Interdisciplinary Center in Interactive Technologies, University of São Paulo (CITI-USP), São Paulo, Brazil

*Address all correspondence to: lidialenzi@alumni.usp.br

## References


[1] Weiser M. The Computer for the 21st Century. Scientific American. 1991; 265(03):94-104.

[2] Espinilla M, Villarreal V, McChesney I. Ubiquitous Computing and Ambient Intelligence—UCAmI. Sensors. 2019;19(18): 4034. DOI:10.3390/s19184034

[3] Lyytinen K, Yoo Y. Issues and Challenges in Ubiquitous Computing. Communications of the ACM. 2002; 45(12):63-65.

[4] da Silva Moreira ALC, Archanjo JM, de Deus Lopes R . Potentialities of the Internet of Things in the Health Area in Brazil. In: Santos H, Pereira G, Budde M, Lopes S, Nikolic P, editors. Science and Technologies for Smart Cities. SmartCity 360 2019. Lecture Notes of the Institute for Computer Sciences, Social Informatics and Telecommunications Engineering, vol 323. Springer, Cham. DOI: 10.1007/978-3-030-51005-3_3

[5] International Longevity Centre Brazil - ILC-Brazil. Active Ageing: A Policy Framework in Response to the Longevity Revolution. Rio de Janeiro: ILC-Brazil; 2015.

[6] United Nations, Department of Economic and Social Affairs, Population Division (2020). World Population Ageing 2019 (ST/ESA/SER.A/444).

[7] Denton FT, Spencer BG. Chronic Health Conditions: Changing Prevalence in an Aging Population and Some Implications for the Delivery of Health Care Services. SEDAP. 2010; 29(01):11-21. DOI: 10.1017/S0714980809990390

[8] Steinmann L, Telser H, Zweifel PS. Aging and Future Healthcare Expenditure: A Consistent Approach. Forum for Health Economics & Policy. 2007; 10(2).

[9] World Health Organization. Adherence to long-term therapies. Evidence for action. Geneva; WHO: Library Cataloguing-in-Publication Data; 2003.

[10] Jack K, McLean SM, Moffett JK, Gardiner E. Barriers to treatment adherence in physiotherapy outpatient clinics: a systematic review. Man Ther. 2010;15(3):220-228. DOI: 10.1016/j.math.2009.12.004

[11] Patel KK, Patel SM. Internet of Things-IOT: Definition, Characteristics, Architecture, Enabling Technologies, Application & Future Challenges. International Journal of Engineering Science and Computing. 2016;6(5): 6122-6131. DOI 10.4010/2016.1482

[12] Infocomm Media Development Authority. Internet of Things (IoT). Singapore. Available from: https://www.imda.gov.sg/-/media/imda/files/industry-development/infrastructure/technology/internetofthings.pdf%3Fla%3Den

[13] Statista. Forecast number of IoT connected objects worldwide from 2018 to 2025, by type. Germany. Available from: https://www.statista.com/statistics/976079/number-of-iot-connected-objects-worldwide-by-type/

[14] Statista. Prognosis of worldwide spending on the Internet of Things (IoT) from 2018 to 2023. Germany. Available from: https://www.statista.com/statistics/668996/worldwide-expenditures-for-the-internet-of-things/

[15] Deloitte Centre for Health Solutions. Medtech and the Internet of Medical Things: How connected medical devices are transforming health care. United Kingdom, 2028.

[16] Statista. Percentage of adults worldwide who currently used a

connected health device or tool to manage their health for select reasons as of 2018. Germany. Available from: https://www.statista.com/statistics/917185/connected-device-usage-reasons-worldwide/

[17] Farahani B, Firouzi F, Chang V, Badaroglu M, Constant N, Mankodiya K. Towards fog-driven IoT eHealth: Promises and challenges of IoT in medicine and healthcare. Futur Gener Comput Syst. 2018;78:659-76.

[18] Pratap R, Javaid M, Haleem A, Suman R. Internet of things (IoT) applications to fi ght against COVID-19 pandemic. Diabetes & Metabolic Syndrome: Clinical Research & Reviews. 2020 14:521-524.

[19] Dash SP. The Impact of IoT in Healthcare: Global Technological Change & The Roadmap to a Networked Architecture in India. J Indian Inst Sci [Internet]. 2020;100(4):773-785. DOI: 10.1007/s41745-020-00208-y

Chapter 5

# A Thermal and Energy Aware Framework with Physiological Safety Considerations for Internet of Things in Healthcare and Medical Applications

*Amitabh Mishra*


## Abstract

Healthcare, lifestyle, and medical applications of Internet of Things (IoT) involve the use of wearable technology that employs sensors of various kinds to sense human physiological parameters such as steps walked, body temperature, blood pressure, heart rate and other cardiac parameters. Such sensors and associated actuators can be worn as gadgets, embedded in clothing, worn as patches in contact with the body and could even be implanted inside the body. These sensors are electronic, and any electronic activity during their sensing, processing and wireless transmission is associated with the generation of heat. This dissipated heat can cause discomfort to the subject and has the potential of damaging healthy living tissue and cells. In the proposed work, the author does a performance check on the intrinsic safety aspects of an IoT healthcare network with respect to the functioning of the wireless sensors involved and routing of sensor data samples. The author also suggests an optimized thermal and energy aware framework to address the issue of temperature rise due to processing and data transmission from sensors through signal processing approaches that help in reducing thermal hazards and simultaneously enhancing the network lifetime through energy conservation.


**Keywords:** Internet of Things, Internet of Things in healthcare, wireless sensors, safety considerations in IoT applications, power for wireless devices

## 1. Introduction

All living beings require healthcare and monitoring, and the requirement increases with age. According to the Department of Economic and Social Affairs of the United Nations Secretariat, the elderly population (persons of age 60 years and over) in the world in 2020 was 1049 million and is projected to be 1,198 million in 2025, or 15% of world population [1]. Healthcare is expensive and the treatment and its management require a lot of data collection. Occurrence of pandemics amplifies healthcare requirement for living beings of all ages, and more so for geriatric subjects, pressurizing the healthcare systems. Medical cost trends are increasing all over the world for multiple reasons and are expected to maintain an upward trend in the future, irrespective of the healthcare models used by the different countries in the world. According to a study by PricewaterhouseCoopers Health Research Institute, there will be a 7% medical cost trend in 2021, a percent above the trend in 2020 [2]. A study on healthcare spending by Peterson foundation reported that during 2019, the spending was close to $3.8 trillion, or $11,582 per person in the U.S. These costs are expected to climb to $6.2 trillion—roughly $18,000 per person by 2028 [3].

## 2. Requirement of a new system for ubiquitous health monitoring

In most healthcare systems, rocketing expenses, insufficient staffing, medical inaccuracies, and the incapability of the patient to get to a hospital in time are adding to the workload of the already overloaded existing healthcare provisions. Vital parameters for living subjects often require monitoring that needs appropriate sensors. Use of wires for sensor data transfer requires the patient to be either stationary or that sensors, electronics, wires, and human-machine interface (HMI) unit, all move with the subject. A wired monitoring system impacts the mobility of the subjects. It is also a major inconvenience to patients if they must visit hospitals every time for getting the readings of vital parameters taken. Such monitoring done only during hospital visits is not continuous, gives the healthcare professionals merely a snapshot of the patient's health parameters for a short time window, and is hence neither efficient nor perfectly reliable. Mobility of geriatric patients using such a wired system could be even more difficult. Quite a few times, the subject does not need to be confined to a bed and the health parameters still need to be monitored. The traditional healthcare monitoring sensor system designed using wired connections is cumbersome and impracticable for such applications.

There is a strong need for ubiquitous and pervasive monitoring of physiological, biochemical, and physical parameters in any environment without activity constraint and behavior alteration for managing patients with chronic ailments and geriatric care. Other important use cases could include general monitoring of well-being of any subject, performance evaluation of sportspersons and deployed soldiers and other applications involving travel and distant patients.

With recent advances in wireless technologies, it is possible to get rid of the wires and relay the data from the sensors to the HMI unit over wireless links, often via multiple hops across wireless transceivers built into the IoT sensors, thus creating an Internet of Things - Healthcare Sensor Network (IoT-HSN) that can exist in or around the subject's body.

To address the design requirements of an IoT-HSN, the technical issues that need to be focused on include the necessity for wearable or implantable devices with better sensor design, power source miniaturization with possible energy harvesting, biocompatibility, Micro Electro-mechanical Systems (MEMS) integration, low power wireless communication, secure data transmission and seamless incorporation with smart therapeutic schemes. The design would also benefit from redundancy and complementary sources of data to boost the information content and lessen systematic and random errors in sensor data. What is even more important is that such a system must do this inexpensively.

Non-intrusive, ambulatory, continuous, yet economical health monitoring systems using IoT-HSNs are now being developed to achieve a better and complete picture of health diagnosis and reduce the cost of healthcare. In this approach, multiple miniature, battery-powered, networked wireless sensor devices can be attached to or implanted inside the subject's body. These devices sense and collect data on subject's vital signs and transmit the data wirelessly to a central device implemented in a personal digital assistant (PDA) or a smartphone that collects and sends the data to a base station over an external network making them available to healthcare personnel for further assessment and analysis. The system obviates the need for wires that restrict the subject's movement and confine the subject, thus making ubiquitous but unobtrusive monitoring possible.

While IoT-HSNs are extremely useful and the need of the day, human tissue can be harmed by the heat produced by the electronic circuitry for the sensor node and antenna. This paper tries to address this issue related to IoT-HSNs in a novel way at the physical and data-link layer level.

## 3. Primary motivation for the development of IoT-HSNs

For prevention and complex intermediation, clinical practice relies heavily on early, truthful, and thorough diagnosis supported by tight scrutinizing of the results. To obtain qualitative and quantitative data for physiological parameters for living beings, a variety of sensors have traditionally been in use. These sensors need to convey their data to an HMI unit that can collect, analyze, and display the data in a variety of formats for use by healthcare personnel and store the data for future use. Traditionally, such data is relayed over wires to the HMI unit. The complexity of such a system increases with the number of physiological parameters being monitored. However, for the most part, this practice depends on a sequence of snapshots of physiological, bio-mechanical, and biochemical data which might not capture transient abnormalities reliably. An objective determination of a patient's recovery after diagnosis can be tricky due to the episodic and subjective nature of outpatient clinic assessment.

Vital signs monitoring systems for hospital ward-based patients have a propensity to be intensive on labor as they involve manual measurement and documentation, which also makes them prone to human error. Such systems restrict patient movement which might be redundant in several cases and can be benefited immensely by using wireless sensors. Automation of this process using wireless sensors with the capacity to pervasively observe patients wherever they are, not just on a hospital bed, is suitable to the patient as well as the healthcare provider.

Acute as well as chronic disease management through clinical medicine, health monitoring and healthcare delivery need to involve home and community settings and require radical changes in system design. Close monitoring of some patients needs to be made possible with safe early discharge without hospitalization being necessary, also reducing the cost for the patient and improving hospital bed availability. The pandemic has already proved that availability of hospital beds and their management can be extraordinarily challenging and critical at times.

### 3.1 A special case: Elderly patients

There are rapid changes happening in the social and economic structure of our society connected to demographic variations associated with increase in vulnerable aging population living alone, a sizable part of which constitutes the high-risk group that would benefit immensely by regular and non-intrusive healthcare monitoring. The volume of this group is set to expand, along with its prospective need upon healthcare resources because people in industrialized countries are living longer than ever before and average life expectancy has improved to more than 65 years [4].

The incapacity of the elderly residents to get medical assistance early enough for simple and treatable conditions may lead to substantial morbidity. Inclement and extreme weather conditions and the fact that they live alone could be two major factors responsible for delayed medical intervention that could make things worse.

It is an additional consideration if they live in rural areas. There is an acute need for unobtrusive monitoring of such patients in their home environment in any weather for earlier detection of any worsening in their condition, so that they can be promptly treated, thus reducing the necessity for hospital admission, related morbidity and even chances of mortality.

### 3.2 Novel trends involving lifestyle modifications

In recent times, the focus of healthcare also altered towards the general health and wellbeing of the populace rather than just the supervision of disease advancement or the effectiveness of therapeutic processes. Several healthy people actively monitor their health parameters because of increasing awareness towards healthy living these days. This is required for patients as well. Certain critical health-related events might not occur in the time window when the patient is in front of healthcare professionals. Such events could be missed, make a difference to the diagnosis and treatment, and thus create room for error. Therefore, several patients require health monitoring although they do not have to live in a hospital for this purpose.

Health is defined as "a state of comprehensive physical, mental and social well-being and not simply the non-existence of illness of infirmity" by the World Health Organization (WHO) [5]. Blocking disease through campaign of healthy lifestyle choice is a prospective cost-effective methodology to address contemporary healthcare risks [6]. The healthcare approach is shifting towards watching lifestyle behaviors and intervening when essential.

Selections such as smoking and alcohol, diet, sleep, physical activity, have all been linked with numerous medical conditions. The cardiovascular disease is one of the most documented illnesses related to lifestyle choices today [7]. Undesirable lifestyles that lead to chronic conditions need to be advocated against, in favor of promotion of healthy living with prevention and early intervention of ailments. There is plenty of evidence to link inactivity with poor physical condition which is why physical activity monitors are commonly available today and are still evolving for better efficiency [8].

The user-friendly software that comes with these activity monitor sensors is true value addition because it permits customized activity targets to be established, and progress towards those targets to be presented at any time or archived and examined later. The software can help with weight monitoring and management as well as diet tracking. Such monitors have demonstrated that they enhance quality of life as much as expensive, overseen workout programs [9].

### 3.3 Some prominent challenges for IoT-HSN applications

Anomalies of heart rhythm (arrhythmias) are frequently confronted in clinical practice, affecting almost 4% of the populace beyond the age of 60, rising with age to roughly 9% in people above 80 [10]. Heart failure affects up to 10% of patients who have attained an age of 65 years [11]. Early symptoms of atrial fibrillation arrhythmias include fatigue and palpitations, and often lead to the patient seeking medical advice. Averting the longer-term issues of tachycardia (rapid heart rate induced) involving cardiomyopathy (expansion of the heart causing pump failure) and stroke in such patients becomes crucial. Prospective bleeding problems caused by anticoagulant medication affect an escalation in mortality in this geriatric patient cluster, in addition to other risk factors [12]. Continuous and pervasive monitoring of heart rate is desirable for several patients and the elderly.

One of the principal vital signs, the systemic arterial pressure (ART) outcomes from the pressure exerted by the circulating blood in the large arteries and is then measured within large arteries in the systemic circulation in mmHg units. The parameter is dependent upon cardiac output and total peripheral resistance and its value varies with each heartbeat in accordance with the pumping action of the heart. All levels of ART exert some systematic stress on the arterial walls. Arterial pressure directly relates to cardiac output, arterial elasticity, and peripheral vascular resistance [13]. It is vital for the subject's body to be capable of adjusting to acute changes in arterial pressure and for the subject to obtain medical therapy or lifestyle modifications for chronic variations. Arterial pressure regulation is required to sustain a sufficiently high pressure that permits appropriate perfusion of body organs and tissue; but not high enough to cause harm. The connected medical condition is known as essential hypertension and is seen in roughly 95% of patients with hypertension [14, 15]. Treating hypertension is crucial because it can cause cerebral, cardiac, and renal problems. As it is a key parameter connected to the cardiac condition of the subject, the author decided to choose the analysis of this parameter as a representative of vital signs for the present work while the author dealt with data for several other equally important parameters.

Atrial fibrillation is known to have several associated complications such as hypertension or high blood pressure. High blood pressure is known to affect nearly one billion persons globally [16] and can relate to cardiac problems. Early identification of hypertension is vital, but its monitoring can be labor-intensive and might involve several clinic visits.

## 4. Technological advancements in favor of wireless health monitoring

The technology for new biological sensing modalities has started emerging and it aims at basically transforming the way we utilize bio-measurements in a truly customized monitoring platform that is smart and context-aware, yet imperceptible. An IoT healthcare sensor network (IoT-HSN) consists of one or more wireless sensor devices positioned on, in, or around the human body. The sensor devices sense and collect data from the human body and then transmit the data to a central device, called a Coordinating Sink Station (CSS) or simply sink, that can be implemented as an application in a smartphone or PDA. After collecting all information, this sink then forwards the data to the medical workers through external networks.

Thus, the idea behind an IoT-HSN is to perform the monitoring of human well-being in a "ubiquitous" and "pervasive" way keeping an eye on physiological, biochemical, and physical parameters in any environment – home or hospital, without constraint of activity [17, 18]. This idea is rapidly converting to reality with the key innovations in sensors, processor miniaturization, and wireless technologies for transmission of sensor data [19, 20].

*Telestethoscopy* is one such application in which electronic stethoscopes created by adding a capacitive diaphragm sensor with microphones and piezoelectric crystals [21] are making remote cardiopulmonary examination of patients in their home environments possible [22].

Innovations in crucial areas such as miniaturization of power supply, enhanced battery time, lowered energy intake, and power scavenging are vital to the design of such systems and are fast becoming a reality [23]. Use of customized wireless sensor network (WSN) technology for creating pervasive healthcare systems will permit access to truthful medical information irrespective of place and time and will go a long way in improving the quality of healthcare services.

Due to the restricted bandwidth and power constraints in an IoT-HSN, the optimality of conventional method of data acquisition followed by post transmission digital conversion and signal processing is questionable. While it requires resources, bio-inspired local processing at the sensor front-end prior to transmission, combined with behavior profiling, pattern recognition, and machine learning can yield highly optimized bio-monitoring systems.

## 5. Why IoT-HSNs are different

An IoT-HSN has more challenges than other wireless sensor networks because of several reasons, the most important of them all being the involvement of living subjects. The various design considerations for IoT-HSNs involve size, cost, reliability, data privacy, security, and intrinsic safety of the subject. This paper tries to address some of these issues concerning the intrinsic safety aspect of IoT-HSN design and the energy efficiency of an IoT-HSN.

WSN technology has benefited by miniaturization and cost reduction in creating sensors with computers and wireless transmission capability that are smaller than the size of the pin head [24–27]. Sensors that can be combined, run on low power, communicate over wireless links, and self-organize into a network have been used in oil and petroleum exploration and industry [25, 28, 29], structural monitoring [29], habitat monitoring [30] and smart homes [31, 32]. Security and scalability of IoT applications and services could also be an issue as addressed in this project aimed at building a Smart Independent Living for Elders (SMILE) home [33, 34] that the author is a part of.

However, the equipment used for these applications cannot address the specific challenges related to human body monitoring. The human body comprises of a complex internal ecosystem that reacts to and interacts with its external environment while staying distinct and self-contained. Hence, although an IoT-HSN is similar in operation to a regular WSN, it comes with an additional set of new challenges. It involves a smaller scale network (made up of miniature sensor nodes each having a small processor, wireless transceiver, and power) that requires a different type and frequency of monitoring and is capable of seamlessly integrating with home, office, and hospital environments.

The IoT-HSN sensor node guarantees the perfect gathering of data from the transducer element used, performs low level local processing of transducer data, and then transmits this data to a Local Processing and Coordinating Sink Station Unit (CSS). The data from all the sensors is collected by the CSS by this method, processed further, fused, and transmitted wirelessly to a central monitoring server [35].

As pointed out earlier, while some of the challenges faced are common to IoT-HSNs and WSNs, there are intrinsic variations between the two, which require special consideration in case of IoT-HSNs. Some of these sensors need to be implanted inside living human tissue. The power source for IoT-HSNs, if exhaustible and hence with finite lifetime, could be inaccessible and difficult to replace in an implantable setting. Energy is more difficult to supply, hence lower the requirement (with options of energy scavenging), the better. Loss of data in an IoT-HSN can be intolerable and may necessitate extra actions to guarantee quality of service (QoS) and real-time data examination capabilities. Human body is capable of movement, so an IoT-HSN is a mobile and dynamically changing network. Motion artifact is a major challenge in IoT-HSNs. Early detection of adverse events is vital in IoT-HSNs because failure of human tissue cannot be reversed. High level security for wireless data transfer is necessary to safeguard patient information and privacy. All these factors change the sensing modalities for IoT-HSN.

## 6. The temperature rise problem: prior work

IoT-HSN sensor nodes could be located on, around, or inside the human body, with each dissipating some part of its energy consumed as heat and causing temperature increase in its locality. Signals carrying sensor data need to travel through tissue (bones, flesh, and fluids). The longer a node works and transmits/receives data, the more energy is dissipated and converted into heat. Nodes not transmitting or in sleep with low power might not produce significant heat. However, continuous node operation over a period generates heat that cannot be ignored. When implanted nodes are being considered, this generates even higher concern. To balance the heat, the human body has a thermoregulatory system. If the rate at which heat is generated is greater than the rate of working of the thermoregulatory mechanism, the temperature rise can harm the human tissue that absorbs the heat. The temperature rise directly influences human safety and health adversely, as explored in [36, 37].

Due to the lossy nature of the human body, the sensor data might hop through intermediate sensor nodes before reaching the sink node instead of being communicated in a single hop. Natarajan et al. [38] attempted to compare the trustworthiness of single-hop and multi-hop network topologies.

The operation of node circuitry and radiation due to transmission from the antenna produce and discharge heat to the node's surroundings, which can be injurious to the subject's body cells beyond a safety threshold. Specific absorption rate (SAR) is a standard quantity that shows the power dissipated per unit mass of tissue. It is a well-known parameter regarding the electromagnetic safety towards the human body and is defined as a measure of the rate at which energy is absorbed by the body when exposed to a radio frequency electromagnetic field, expressed in W/kg [39]. For near-field exposures the upper bound of SAR is 1.6 W/kg for some tissue averaged over a gram according to the Federal Communications Commission (FCC) standard in the United States and is 2.0 W/kg for 10 g of tissue according to the International Commission on Non-Ionizing Radiation Protection (ICNIRP). These SAR values can be translated into temperature rise [36], with the maximum permissible temperature rise in the human head and brain being 0.31°C and 0.13°C (FCC) and 0.60°C and 0.25°C (ICNIRP). The report in [37] also suggests that a temperature increase of 0.1°C is sufficient to cause intense thermoregulatory responses in the human body.

According to a survey on thermal effects of bioimplants by Lazzi [39], the electromagnetic fields induced in the human body and the power dissipated by the implanted sensor nodes are the two main sources of temperature rise. The power dissipation is from three sources: caused by the stimulating electrodes, the implanted telemetry coil, and the implanted microchip.

The in-vitro (implanted) sensor nodes can transmit and receive the data only through a wireless system. The SAR measures the rate of energy absorption by the body per mass of tissue upon exposure to a radio frequency electromagnetic field. It is a standard parameter connected to the electromagnetic safety regarding the human body, expressed in W/kg. According to IEEE standards the acceptable value of SAR is 1.6 W/kg averaged over a gram of tissue and is used for cellular phones by the FCC.

A better hardware design with node and antenna running on lower power can reduce the heating effects. Also, a well-designed network routing protocol could reduce the bioeffects. This work tries to reduce these heating effects even further by reducing the amount of transmission, while trying to preserve the integrity and accuracy of the data within low limits of error as a trade-off of the suggested framework.

### 6.1 Routing based approaches towards solving the temperature rise problem

As briefly touched upon in the previous section, one of the topmost concerns in the design of IoT-HSNs involve monitoring the heat generated because of operation of sensor network nodes. Electronic activity in the sensor circuitry and antenna radiation dissipates as heat. Power is dissipated by the implanted sensor node electrodes, microchips and the electromagnetic fields induced in the human body from telemetry coils as heat which can cause harm to healthy cells and tissue [40, 41]. For burst data operations that do not last long, such heat can be overlooked. However, when the node is operating continuously, transmitting, and receiving data over a considerable period, the heat generated by the node cannot be neglected. This concern becomes even bigger when dealing with in vivo sensor nodes (i.e., implanted inside the human body). The human body has a thermoregulatory mechanism to balance the heat around the body. However, when the heat received rate is larger than the thermoregulatory mechanism rate, the temperature will rise and, in turn, damage the human tissue.

Routing overheads have a potential to cause additional heat damage. Also, extra energy might be required to implement thermally aware routing algorithms. The challenge is complicated by the fact that the heat and energy consumption, both these factors need to be lowered, because the sensor nodes run on the limited power resource of batteries, while the network throughput needs to be maximized. A trade-off needs to be reached in the design to address these diverse requirements.

There are three types of routing used on IoT-HSN protocols. First, proactive routing where each node has information about the neighbor nodes. Second, reactive routing where the node explores the information about the neighbors when there is a packet to be sent. Third, a hybrid which combines the benefits of two methods (e.g., protocols that use proactive in setup phase and reactive in data transmission phase).

Some approaches to reduce the risks of this heat damage involve designing routing protocols for IoT-HSNs that include temperature into the routing metric to decrease the heat.

The challenges related to IoT-HSNs have been proposed to be addressed through numerous routing protocols. Some approaches have tried to tackle the issue of extreme and dynamic path loss observed in intra IoT-HSN communication caused by postural movement of the subject's body. The routing scheme by Quwaider and Biswas [42] proposes division of the sensor field combined with store and flood mechanism to route the sensor data towards CSS. Their work in [43] uses a store and forward approach for a delay tolerant intra IoT-HSN communication.

The proposal in [44] uses a field partitioning with store and forward like in [42] based on if or not the sensors have a clear line of sight for communication. The storage of packets in these works makes the routing non real time making the scheme impractical for vital medical applications. The proposals do not take the heterogeneous nature of IoT-HSN data and the thermal effects into account.

In [45] the routing protocol uses a Temperature Aware Routing Algorithm (TARA) to reduce the thermal effects IoT-HSN operation by estimating the temperature rise in neighboring nodes to avoid hotspot nodes. The trade-off involves a delay in routing sensor data packets and additional energy requirement. The Least Temperature Rise (LTR) algorithm [46] tries to address this limitation by associating a hop-count with each data packet and use it for deciding to discard the packet if the hop count reaches a limiting value. The trade-off in this case is poor packet delivery ratio. Adaptive Least Temperature Rise (ALTR) algorithm proposed in [47] is also a thermal aware scheme that uses shortest hops to route packets instead of dropping them. Least Total Route Temperature (LTRT) algorithm [48] observes the temperature across the entire route instead of individual nodes or hop-count for routing decisions. None of these schemes consider the dynamic intra network path loss or the QoS parameters of heterogeneous IoT-HSN data, making their utility questionable.

Djenouri and Balasingham [49] propose to divide the vital sign data into four categories based on data criticality, thus allowing for delay in some parameters and employ two sinks for all data. The latter feature increases network traffic. Razzaque et al. [50] tried to improvise on [49] by using multi-hop transmission to meet QoS requirements of data packets but their algorithm performs poorly on data packet delivery. QoS aware routing used in two proposals by Khan et al. [51, 52] involves classification approaches that are variants of [49]. None of the QoS-aware routing schemes take inter IoT-HSNs communication, path loss or temperature issues into account.

Monowar et al. [53] and Bangash et al. [54] claim to propose QoS as well as thermal aware routing schemes for intra IoT-HSNs. Both schemes classify the sensor data as in [49, 50]. Monowar et al. [53] propose to send multiple copies of data to counter delay issues. This generates redundant additional network traffic, causes congestion and packet drops despite higher energy requirements and rise in temperature while neglecting the dynamic path loss. The proposal by Bangash et al. [54] performs better on these factors but fails to address the issue of reliable, timely delivery of critical data.

Critical Data Routing (CDR) proposed in [55] classifies data into critical and noncritical categories while considering path loss, temperature rise and QoS with decent performance. However, the scheme could benefit by considering additional measures for conserving network energy, which it does not focus on.

The approach in [56] suggests a Media Access Control (MAC) protocol that resorts to shortest hop routing of sensor data packets based on hop counts using a duty cycle decided upon by using the current temperature rise. The duty cycle is calculated using four probability distribution functions- Poisson, Binomial, Log-normal and Laplace. This protocol was chosen by the author for the current article as no other protocol blends thermal awareness with efficient duty cycles. The work uses three models. Of these, the Sensor-Centric Monte-Carlo model (SCMC) involves random sampling from a given finite space [57] while acquiring any temperature rise right from the sensor and not from the surrounding tissues. In the Tissue-Based Fixed Coordinator (TBFC) model, a grid divided control volume of tissue space is considered, like [39, 45, 58] which assumes that the entire IoT-HSN or a major portion of it is within this tissue control volume. The results indicate least packet loss of 30% for Poisson distribution on the duty cycle with the trade off with 80% active nodes that need more energy for IoT-HSN operation. The packet loss was further reduced by enhancing the working of TBFC by adding 1-hop caching mechanism (TBFC-1HC) in which data packets are cached before the node goes to sleep state if the node has not reached its sampling state while the next hop node might be mere one hop from the CSS.

None of these approaches address the issue of improving upon network lifetime. As the approach in [56] provides for best possible compromise for intrinsically safe, thermal and energy aware IoT-HSN design, the author chose on using it for further optimization and improving upon the energy scenario and network operation lifetime.

## 7. Framework for a novel IoT-HSN with energy awareness enhancement on thermal routing model

The author proposes a model which not only addresses temperature rise but is also energy aware and helps in improving network lifetime. For this study, the author used the same IEEE 802.15.4 Wireless based IoT-HSN schematic modeled for [59] to run in the CSS. The 24 channels in the IoT-HSNs were used to mimic relaying of physiological parameters from the subject such as parietal and occipital electroencephalogram (EEG), electroculography, respiratory airflow, oxygen saturation %, heart rate, pacemaker diagnostics, electrocardiogram (ECG), arterial and central venous pressures, respiration rate, thoracic and abdominal resistance, blood pressure and temperature, blood sugar and insulin levels, urine creatinine, nerve conduction, musculature actuator and electromyography (EMG). The study did not involve any human subjects directly because the data utilized were obtained from Physionet [60], a public research database. Of the 24 channels, 3 were used for bioactuators and the remainder were utilized by sensors. **Figure 1** shows the biosensors and bioactuators using an adhoc link to communicate with the coordinating sink station which was connected over an adhoc link with the body area network (BAN) gateway which in turn links the biosensors to a IoT-HSN base station. To demonstrate the in-depth analysis and to evaluate the performance of the thermal and energy aware framework proposed in this article, the author has used the arterial pressure parameter from the 24-channel model in the following sections.

**Figure 1** shows four different 24-channel IoT-HSNs P1 to P4 in the vicinity of each other trying to send data to the base station with their performance possibly affected by radio interference. The channels in the IoT-HSN have 802.15.4 adhoc links to the BAN gateway for data transmission. Subsequently, the data is sent through a router to the base station. IoT-HSN P2 transmits its data to a different wired base station that exists on the same subnet. IoT-HSN P3 attempts to send data to its base station which is in a different subnet and uses a second router for connections. The base station for IoT-HSN P4 is a wireless node linked to the wired network via an access point. As human subjects can have different sizes, the placement distance for biosensors varies in the four IoT-HSNs.

## 7.1 Performance check on intrinsically safe routing models

While assessing the accomplishment of an IoT-HSN, it becomes vital to evaluate the intrinsic safety aspect of the wireless system and the possible risks of damage to healthy human cells and tissues. As pointed out earlier, the heat generated because of dissipation of wireless energy can cause discomfort to the subject and has the capacity to damage healthy human cells and tissues if endured for long times. For instance, the incessant monitoring of peripheral capillary oxygen saturation (SPO2)

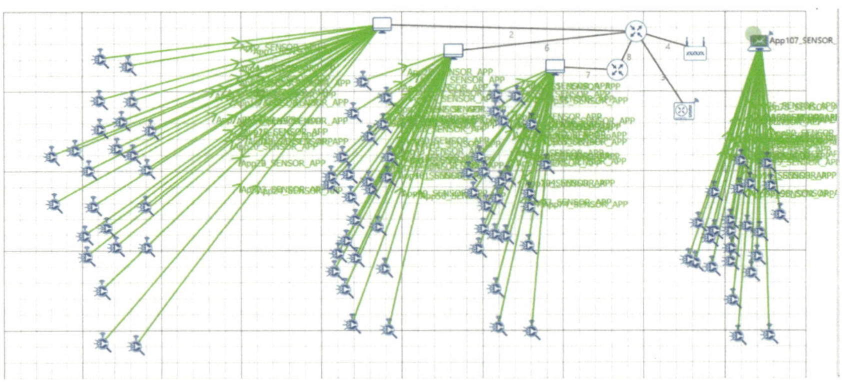

**Figure 1.**
*Four 24-channel IoT-HSNs in action.*

levels using a pulse oximeter for over 8 hours would cause a rise of temperature of 43 degrees Centigrade and is hence deemed risky as it could cause burns [61]. The detrimental effects of such sensor radiation caused heating can be evaluated by applying Penne's bio-heat Equation [62] that offers the heat transfer relationship between the temperature of blood vessels and the tissue surrounding the vessels. IoT-HSNs can follow temperature-aware routing algorithms [63–65] that consider parameters like antenna radiation and the ensuing power dissipation as temperature rise in the surrounding tissue and make routing decisions to minimize the generation of heat. Combined with an efficient MAC protocol, the thermal-aware routing algorithms can be used for generating transmission and sleep duty cycles that allow a reduced rise in temperature than individual schemes [56]. Although the outcomes in [56] are improved over the other attempts at temperature-aware routing, the approach does not take into consideration the base network energy requirement and additional energy consumption required for retransmissions of lost sensor data. The author attempted to estimate the implementation of the three models in [56] with regards to energy in a network involving actuator control applications with sensors for Internet of Things Healthcare Sensor Networks. The model in [56] uses up to 25 sensors in its IoT-HSN, which is very close to the author's model involving 24 sensors [59].

All the models considered in the present evaluation study the effect of four probability distributions for network parameters in addition to temperature rise, namely Poisson, Laplace, Binomial and LogNormal. Of the three, the SCMC model is a sensor-centric model that permits a random generation of packets based on a probability distribution while presuming fixed rise and fall in temperatures. A stable solver comprising of a fixed CSS is employed in the TBFC model for a stepped packet generation to offer improved heat performance than the SCMC. The trade-off for the TBFC model is a higher packet loss which is improved in the third model (TBFC-1HC). This modified TBFC model employs 'one-hop caching' in sensors to cache data packets for transmission delays up to their one hop neighbor that is nearest to the CSS. Data packets wait for a clear-to-send signal after which they are transmitted to the CSS.

### 7.2 Performance evaluation on traffic parameters of the model

The thermal aware routing algorithms for reducing the amount of heat generated have a trade-off in the form of loss of packets. The lost packets need to be retransmitted. The author tried to assess the data overhead due to retransmission resulting from packet loss for the four distributions across the three models. The results of the comparison can be seen in **Figure 2** below. It is evident from the results that of the three models, TBFC fared the worst on the retransmission of packets that were dropped, while SCMC was found to be the best. Comparing the retransmission overhead for the distributions, the Poisson distribution had the lowest values while Log-Normal had the highest retransmission overhead among the four distributions. The work has the potential to be extended by including other distributions involving a more realistic human model.

Even if lost transmissions cause additional data traffic due to retransmissions, a data transmission scheme that involves reducing the frequency on transmissions of the sensor data and sending alternate samples as suggested by the author in the next section would effectively cut down the heating effects in the same proportion. Merely skipping alternate samples would reduce the amount of heat generated to half, thereby allowing longer node operation. If the final recreation does not alter the doctor's initial diagnosis, the sample cut rate can be increased, thereby improving the heat performance to three or even four folds of the default.

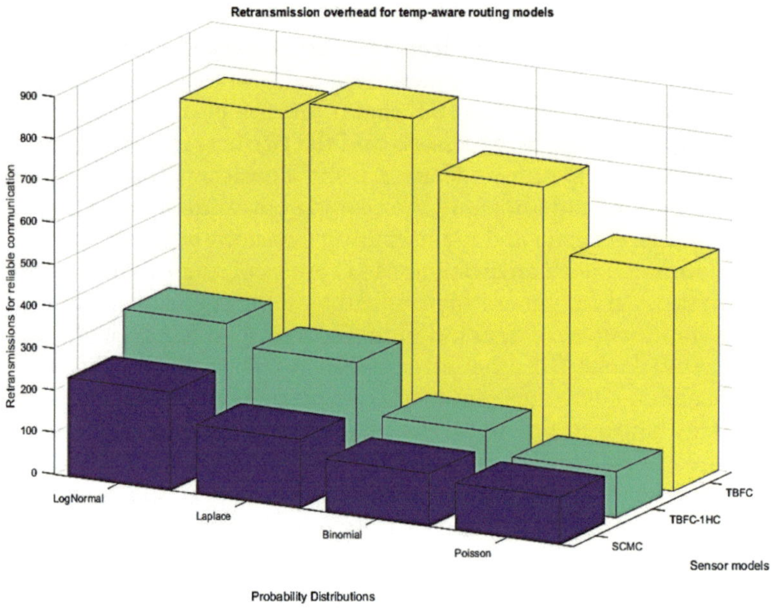


**Figure 2.**
*Number of retransmitted packets in unit time for the three models.*

## 8. Energy saving and network lifetime improvement for IoT-HSNs

A key question related to IoT-HSNs entails the energy-fidelity trade-off. When sensor data is transmitted after processing and transformation, it is expected that the fidelity level of the received data must be acceptable and appropriate to be useful. Any data transformation and transfer need to be done in an energy efficient manner. This requirement advocates for selective processing of collected physiological data samples.

### 8.1 Sample reduction with prediction for energy saving

Another major operation and design issue with IoT-HSNs involves improving the lifetime of sensing for sensor nodes and thus that of the networks. The issue is caused due to the constraints on batteries that need to be small in size and cannot pack a lot of power due to this constraint [66, 67]. The sensor nodes collect data samples and relay them to the CSS at an acceptable rate as dictated by the QoS of the physiological parameter. However, the total number of samples collected and transmitted by the sensor does not take the nature and frequency of variations in the physiological parameter into account by default. In this work, an attempt has been made to address the energy-fidelity trade-off [68] by reducing this data content through signal processing techniques. The approach involved selective exclusion of some sample data from transmission. Prediction techniques were used to recreate the missing samples that were not transmitted. The approach used in this work was different from the dual prediction technique proposed by Mishra et al. [69]. Prediction techniques involve approximations that come with errors but if the error is negligible, the recreated signals can be used for an early diagnosis if not for full diagnosis, while the patient is on the way to hospital.

The fidelity of data and the accuracy of information contained would undoubtedly be better if all the data samples sensed by the physiological sensor were transmitted. Although this sampling approach would satisfy the Nyquist criterion, it would result in transmission of several samples which could be predicted with reasonable accuracy using numerical techniques within some range of error. While such data might not truthfully reflect what continuous monitoring would reveal, the medical personnel would still be helped by early diagnosis, planning or determination on the course of action.

The author first reduced the transmitted samples for each of the parameters in the 24-channel IoT-HSNs to half by skipping transmitting alternate samples and tried predicting the skipped samples at the receiving end by using a simple proportional-integral-derivative (PID) scheme and a more computationally involved prediction using non-linear regression involving an artificial neural network (ANN-NLR). The results are shown for a couple of cycles of prediction for the ART parameter in **Figure 3**. The approximation used in the two prediction strategies generates some error, which is still not too big to alter the characteristics of the ART signal appreciably. This error is shown in **Figure 4**.

The amount of data was reduced by periodically skipping those samples from the original set and predicting the missed samples at the receiving end. The bulk of data marked for transmission could be further used by delta encoding to pack more amount of data in every transmission [59].

Sample data sets for the 24-channel IoT-HSN involving critical physiological parameters such as ECG, central venous pressure, pulmonary artery pressure and arterial pressure signals obtained from Physionet [60], were used as the source to progressively cut down samples and create four different subsets of the original sets like the approach used in [70]. For the sample analysis and graphical evaluation, the programs were written in in MATLAB r2020 [71].

Alternate samples of each of the original sample sets were used to create the first subset, every third sample was picked up to create the second, every fourth sample for the third set and every fifth sample for the fourth set. Thus, the sample sizes of these sets were half, one-third, one-fourth, and one-fifth of the original, respectively. The four sets were transmitted and recreation of original by a variety of numerical interpolation algorithms was attempted at the receiving end. The reconstructed sets were compared with the original set of samples with all samples intact, and the error was calculated. **Figure 5** shows the results of the recreation for

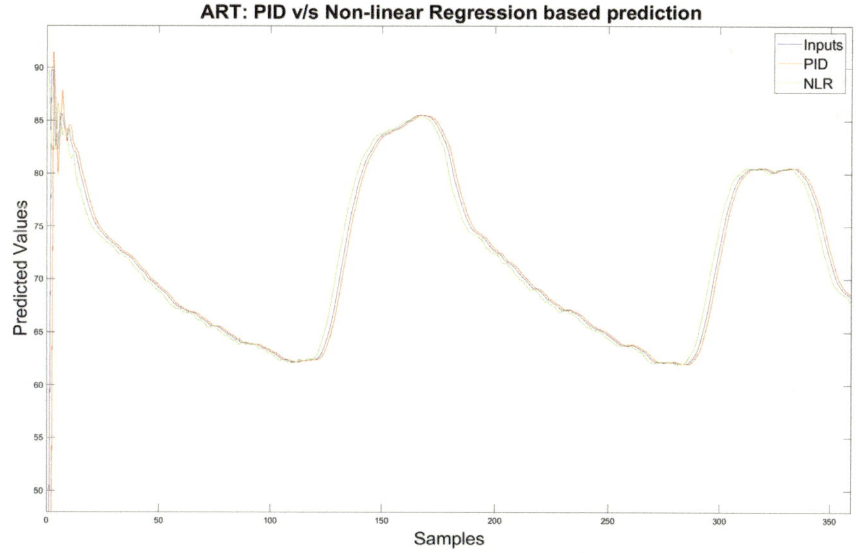


**Figure 3.**
*Plots of comparison of prediction performance by PID and ANN-NLR algorithms for arterial pressure.*

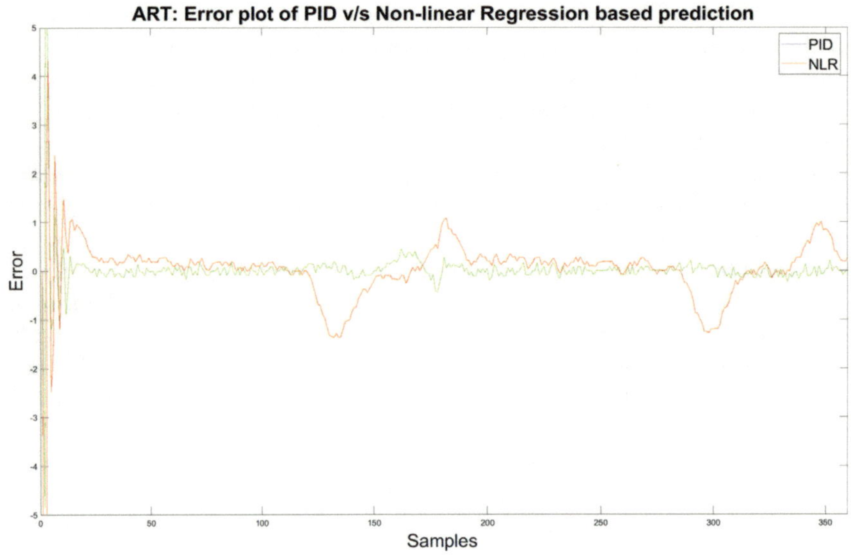


**Figure 4.**
*Error plots of comparison of prediction performance by PID and ANN-NLR algorithms for arterial pressure.*

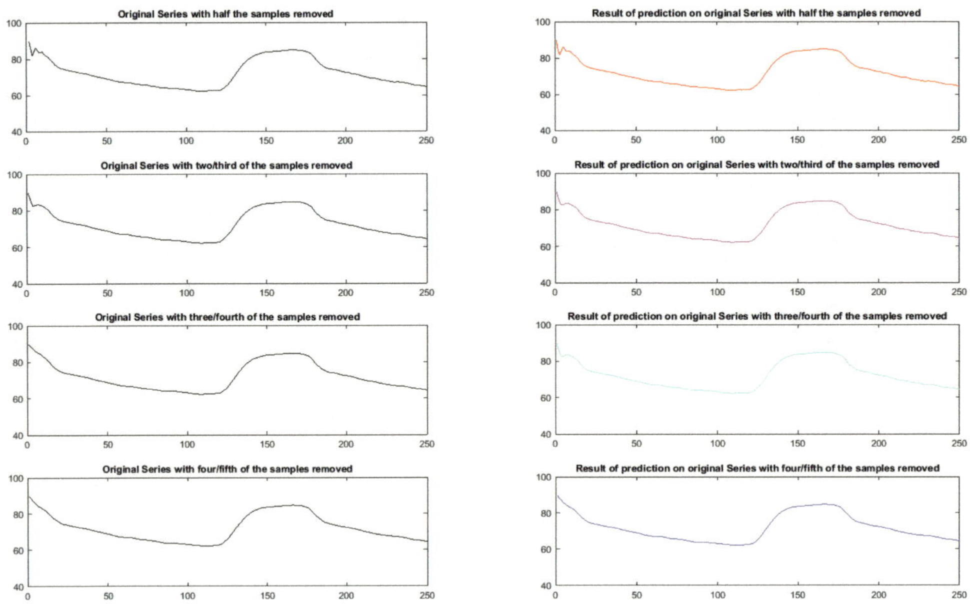


**Figure 5.**
*Prediction of skipped samples for four sample elimination rates through pChip.*

the representative ART signal using ANN-NLR prediction after four different rates of sample reductions, with only a few cycles covered for the sake of conciseness. **Figure 6** shows the error in prediction for the four sample reductions. A similar analysis was also done on the other signals of the 24-channel IoT-HSN with comparable results. **Table 1** shows the particulars of the ART signal used as a representative of the results.

The signals recreated at the receiver using five different interpolation techniques over reduced samples for the arterial pressure parameter were compared with the original full sample sets for error in prediction by interpolation. The results of the

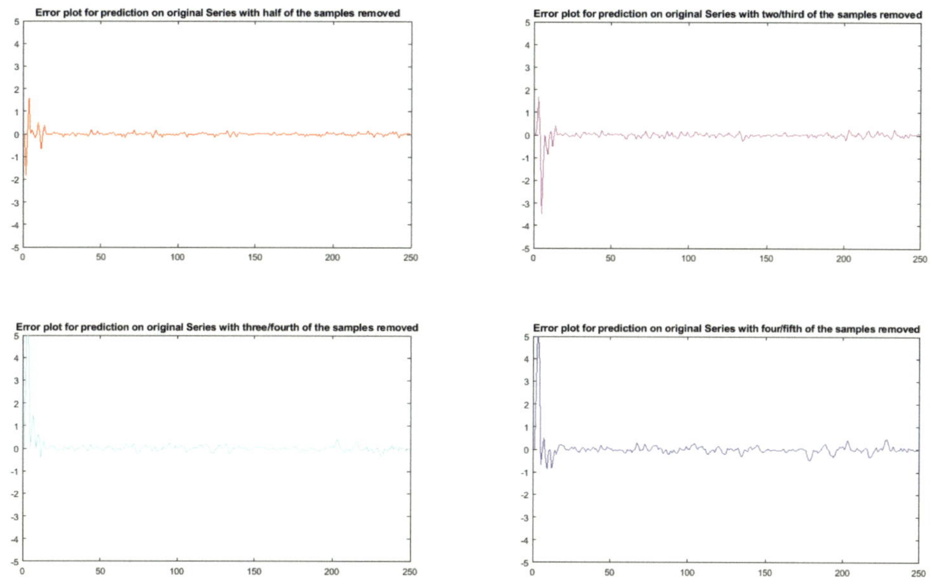


**Figure 6.**
*Error plots for prediction of skipped samples for four sample elimination rates through pChip.*

| Characteristics | Signal | Signal | Signal |
|---|---|---|---|
| | Minimum | Maximum | Span |
| ART (mV) | 52.35 | 89.6935 | 37.343 |

**Table 1.**
*Signal specifications for the arterial pressure vital sign IoT-HSN parameter.*

prediction for the ART signal and the associated error analysis for just the cubic interpolation technique are presented in **Table 2**.

Five numerical interpolation techniques – linear, near, Spline, Pchip and cubic were employed for rebuilding the missing IoT-HSN sample data for the parameters of the 24-channel IoT-HSNs at the receiving end. **Table 3** shows the comparison between the five techniques for the ART parameter.

From **Table 3**, it is evident that the nearest neighbor interpolation algorithm performs the poorest of all but the other four yield lesser error, almost in the same range, with the linear spline interpolation performing better across the sample sets.

Twenty random sets (with 3600 samples transmitted in 10 seconds) of the signals for the 24 channel IoT-HSNs from healthy individuals and patients were employed to assess the performance of the prediction algorithms. The physiological parameters are all different in range, wave-shape, and type of variations. **Figure 3** shows the results of error evaluation after reconstructing the signal using the five interpolation algorithms for one such set of values for the ART signal.

The first column in **Figure 5** illustrates the signal sets with successive reduction in samples. The second column indicates the reconstruction of lost

| 1–6 Sample Reduction | Halved | 1/3rd | 1/4th | 1/5th | 1/6th |
|---|---|---|---|---|---|
| ART (mV) | 1.58 | 1.67 | 6.01 | 5.65 | 1.02 |
| %Error | 0.04 | 0.04 | 0.16 | 0.15 | 1.02 |

**Table 2.**
*Peak error with sample reduction for arterial pressure using cubic interpolation.*

| Reduced | Nearest | Linear | Spline | Pchip | Cubic |
|---|---|---|---|---|---|
| 1/2 | 0.102 | 0.043 | 0.049 | 0.043 | 0.034 |
| 1/3rd | 0.104 | 0.082 | 0.039 | 0.045 | 0.057 |
| 1/4th | 0.12 | 0.166 | 0.158 | 0.161 | 0.162 |
| 1/5th | 0.212 | 0.159 | 0.143 | 0.152 | 0.152 |

**Table 3.**
*Maximum percentage error values for ART from the five numerical interpolation techniques.*

samples for the corresponding row after data reception done using the Pchip interpolation prediction.

## 9. Considerations for battery usage in IoT-HSNs

A key requirement of IoT-HSNs is low power wireless which in turn makes signal detection difficult. Low power wireless is required, which makes signal detection more challenging. Common and proven technologies such as Bluetooth, ZigBee, General Packet Radio Service (GPRS) and Wireless Local Area Network (WLAN) might not offer good and optimal solutions to the low power requirement problem.

The growing miniaturization and cost drop on IoT-HSN sensors, circuits and wireless communication electronics is establishing new opportunities for wireless sensor networks in wearable applications. Nevertheless, for sensors to be untethered, the design needs to use wireless communication between nodes along with wireless powering of sensors. This requirement is fulfilled by batteries in most of the portable electronic devices, making them an obvious answer for IoT-HSN wireless applications. However, the batteries have a finite life and require to be replaced or recharged. This limitation presents a cost and convenience penalty which is undesirable in wireless applications including IoT-HSN while the market for such applications and demands grows. One possible solution to this problem involves harvesting energy from the environment for recharging of power sources. Energy scavenging from motion (vibration) and thermal (body heat) sources offer some options for recharging mechanisms that are being investigated. While the power demands of many electronic functions including wireless communication are being actively reduced, energy efficiency of power sources remains a problem because IoT-HSN nodes are intended to operate for a long period of time, especially if they are implanted.

### 9.1 Batteries and fuel cells for IoT-HSN sensor nodes

Wireless devices can be powered by primary, or rechargeable batteries. Of these, primary batteries are better in energy densities, shorter in leakage rates and lower in cost. The energy density of Lithium-ion batteries that are most used in electronics, is around 700–1400 J/cc for rechargeables [72], and the figure for primary cells is higher. Batteries used for IoT-HSN applications are preferred to last at least a year. A lifetime of 1 year corresponds to 32 J/micro-watt of average power for an average power requirement of some tens of micro-watts.

Hence, a finite battery-life of some tens of microwatt-years is attainable for a battery under 1 cc. Search for better alternatives is on because such batteries require replacement and have issues related to toxicity, safety, and operating temperature range. While ultracapacitors are drawing rising interest for powering electronics as their energy densities are much higher than those of conventional capacitors, the density still are way lower than those of batteries [73].

Hydrocarbon fuels are known to have very high specific energy in the range of 16 kJ/cc for pure methanol [74] or 31 kJ/cc for iso-octane [75]. For miniature electronics, exhaustible sources of energy that use hydrocarbon fuel of some type are also under review, although primarily for greater power levels. Small, micro-machined and only few inches big external combustion heat engines have been built to provide power for portable electronics that can generate up to 200 micro-watts [76]. Such engines have a disadvantage of moving parts and very high temperatures, and hence fuel cells are also being widely investigated for applications involving low power. In the pure methanol-based device used in [74], the electrochemical reaction of methanol with water after passing through a polymer membrane results in oxidation of methanol producing free electrons and protons and generating high power levels of 195 mW/sq.cm.

Miniature fuel cells for implantable sensors as small as 0.5 mm thick can also utilize energy harvesting to provide inexhaustible power up to 4.4 micro-watts/sq. cm if they use body fluids such as oxygen dissolved in blood and glucose as the fuel source [77]. A crucial challenge for such power sources that needs to be addressed is their operational lifetime.

### 9.2 Challenges related to IoT-HSN power sources

Due to difficulties in changing or recharging batteries in IoT-HSN sensors (some implanted), the management of energy consumption for network longevity and resourceful network operation is an important design consideration. The network design methods utilize a sleep-awake cycle for conserving energy and increasing the network operation time because the power requirement for the communications unit in a sensor node is several orders higher in comparison to the transducer and A/D converter unit in the sensor electronics. The author attempted to assess the lifetime of the proposed IoT-HSN framework created using commercial sensors and power supplies focusing on the period that the sensors would remain powered on.

The sample rate for the ART signals used in the representative evaluation was 360 samples per second with the samples encoded in 8-bits and the more popular 12-bits. The total energy necessary for the operation of a IoT-HSN sensor node varies based on factors such as sleep-awake cycle, inter-sensor distances, the time for which the node stays in a specific mode, as well as a system constant.

Based on Heinzelman's sensor node transceiver model [78], the transmission energy required to transmit a k-bit message to a distance of $d$ can be computed as:

$$E_{Tx}(k,d) = E_{Tx-elec}(k) + E_{Tx-Ampl}(k,d) = E_{Elec} * k + \epsilon k d^2 \tag{1}$$

where,
$E_{Tx-elec}$ is the energy expenditure in the transmission electronics,
$E_{Tx-Ampl}$ is the energy expenditure in the transmission pre-amplifier,
$\epsilon$ is an amplification factor.
$d$ is the communication distance between sensors.
Their model has the below assumptions:

$$\begin{gathered} \epsilon = 100 pJ/bit/m^2 \\ E_{Rx-elec} = E_{Tx-elec} = E_{Elec} \end{gathered} \tag{2}$$

To receive a k bit message, the energy expended in the receiver is

$$E_{Rx}(k) = E_{Rx-elec}(k) = E_{Elec} * k \tag{3}$$

The energy expended in the transceiver electronics for most sensor nodes is identical for transmission and reception circuitry and in a few tens of nJ/bit.

The energy required for transmitting all the samples (and not skipping any of the 360 samples) while continuously operating for a minute was 8.65 mJ when the samples are encoded in 8-bits and 12.97 mJ when the samples are encoded in 12-bits.

The author attempted to assess the life cycles of wireless networks comprising of two commercially available low-power ultra-compact sensor nodes. The minute-long sensor duty cycle comprises of 10 seconds each of transmission and reception succeeded by 40 seconds of sleep. The author used three sensor modes for this evaluation – the Eco [79], Texas Instruments' TI CC3100 [80] in Direct Sequence Spread Spectrum mode (1DSSS) and the TI CC3100 in Orthogonal Frequency Division Multiplexing mode (54OFDM).

The Eco sensor required a current of 16 mA during transmission, 22 mA during reception and mere 2 μA while sleeping.

The TI CC3100 sensor fared fine in the 1DSSS mode while performing amply better in the 54OFDM mode. The author evaluated the performance of these sensors based on three commercially available batteries that supply 3.0–3.6 volts, 0.5A – the CR2032, CR123A, iXTRA and ER34615. **Table 4** summarizes the battery characteristics and the findings for transmission power requirements of the two sensor nodes without any power management applied.

**Table 4** indicates that the innovations in low-power sensor design and battery technology enhance the lifetime of the IoT-HSN network.

If the sample reduction algorithm suggested by the author is used, the sensor, and hence the network lifetime would improve in accordance with the sample chop rate. **Figure 7** shows the network lifetime improvement for the sample chop rates.

### 9.3 Battery life for models using thermal-aware routing

The author also attempted to evaluate the performance of the three thermal aware routing models for network lifetime with the three batteries that were shortlisted and considered by [56]. Of the battery models evaluated, the model based on ECO sensor nodes running on the 19000 mAH ER34615 battery had the best performance for network lifetime without any power management.

The TBFC thermal-aware routing model was found to offer the poorest economy on the battery power in these evaluations as compared to the other two models for the four probabilistic packet distributions. **Figure 8** shows the battery and network lifetime for the model despite retransmissions using the mentioned battery-sensor combination in the number of hours of operation, in conjunction with the details in the tables.

The three models pave a way for a study towards efficient and intrinsically safe, thermal-aware IoT-HSNs for wearable computing. **Figure 9** shows the improvement for the sample chop rate of 3, if the reduction in samples is used with

| Battery → | CR2032 | CR2447 | CR123A | iXTRA | ER34615 |
|---|---|---|---|---|---|
| Sensor Node ↓ | 225 mAH | 1000 mAH | 1550 mAH | 1700 mAH | 19000 mAH |
| ECO (16 mA) | 1.76 | 7.82 | 12.11 | 13.28 | 148.62 |
| TI – DSSS (21 mA) | 1.34 | 5.96 | 9.23 | 10.12 | 113.16 |
| TI – OFDM (9.39 mA) | 2.99 | 13.29 | 20.63 | 22.63 | 252.49 |

**Table 4.**
*Life in days for the different battery models as per their capacities and node power requirements if continuous power drawn.*

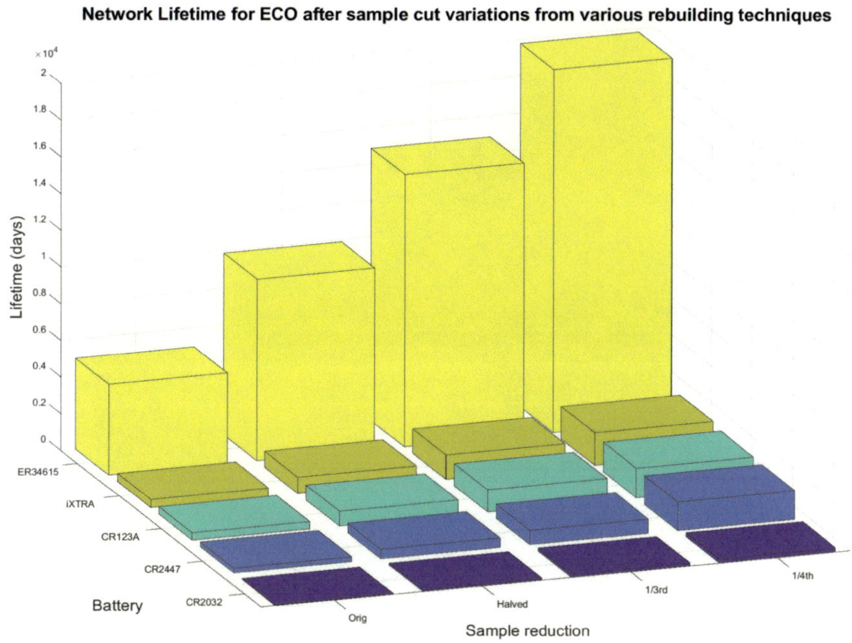


**Figure 7.**
*Comparison plot of battery lives for 5 batteries and four sample rates.*

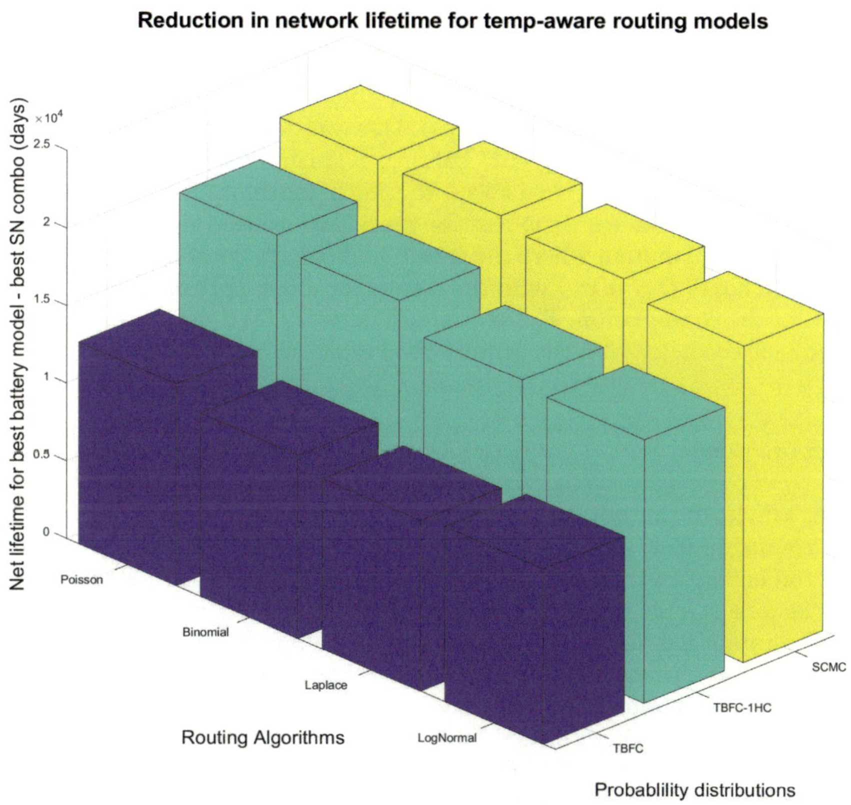


**Figure 8.**
*A comparison of lifetime hours for the three models across four duty cycle distributions.*

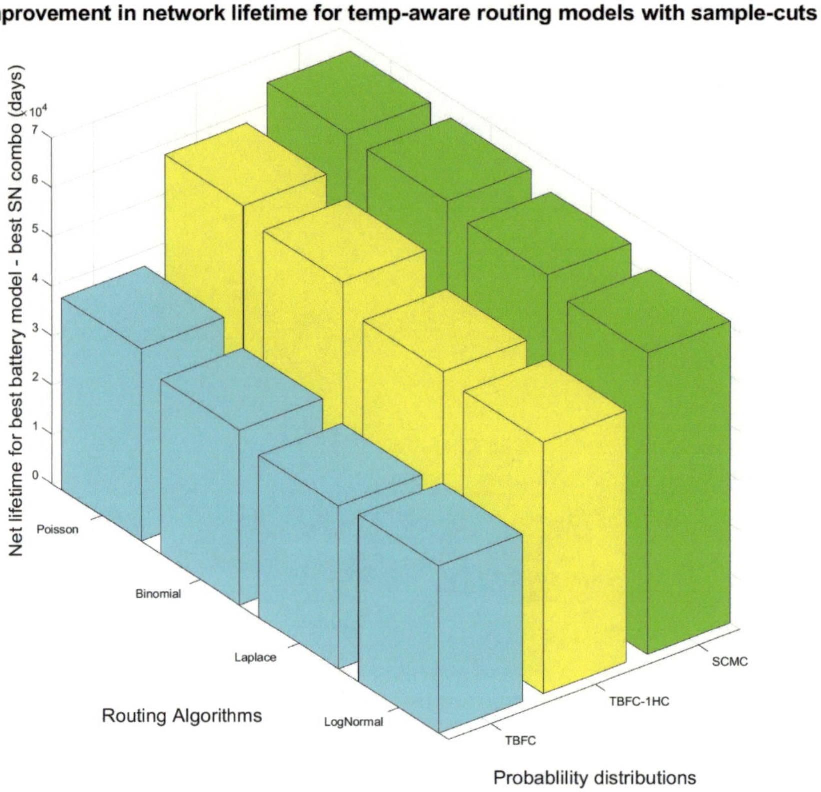


**Figure 9.**
*A comparison of improvement in lifetime hours for the three models for best sensor and battery combination.*

prediction for recreating the original signal. This was done for the ECO sensor when used in conjunction with the ER34615 battery, evaluated over the four probability distributions for the three thermal aware routing algorithms. The author's findings indicated that the results were the best for the SCMC routing algorithm with Poisson sample distribution where the sensor and battery combination lasted for almost 66500 hours (7.6 years) with the results for other distributions not very different for the combination. The sensor and network lifetime are seen to be improved in accordance with the sample chop rate.

## 10. Conclusions

In this article, the author has presented a comprehensive survey of the different types of routing models used for IoT-HSN data. The author has also proposed a thermal and energy aware model that enhances the lifetime of IoT-HSN for intra-network as well as inter-network traffic and evaluated the performance of the model. The author has also demonstrated energy savings by reduction in transmission using a linear elimination algorithm and recreating the missed data at the receiver using a variety of techniques involving a variety of interpolation techniques and prediction using PID and NLR-ANN with very low error values. The savings shown from the model and the enhancement of network lifetime have been demonstrated in quantified as well as graphical forms.

While the basic factors of the network look good for employing energy optimization in IoT-HSN applications, the dynamic execution of the proposed model needs to be studied in better detail for real life HSN applications. An extension of this work could focus on a clinical implementation covering several vital parameters with varying rates of change in them. More possibilities could emerge if the model is tested on well-founded and strong applications such as pacemakers, insulin monitors or movement sensors and prosthetic control.

This article opens the arena for further probing of thermal, QoS and energy aware design of micro-hardware for wearables and implantable bionics. The thermal and energy aware model offers an encouraging prospect to be selected as a design standard for IoT-HSN applications whereas none exists at this time.

## Acknowledgements

The author expresses his heartfelt thanks towards Physionet [60] and the doctors from the Department of Pediatric Cardiology at Cincinnati Children's Hospital for their judgment on the tolerable drop in volume of physiological data for competent supervision of vital body parameters for human subjects.

This work was not supported by any funding whatsoever from any sources.

## Conflict of interest

The author declares no conflict of interest.

## Abbreviations

| | |
|---|---|
| IoT | Internet of Things |
| HMI | Human-machine interface |
| IoT-HSN | IoT healthcare sensor network |
| MEMS | Micro Electro-mechanical Systems |
| PDA | Personal digital assistant |
| WHO | World Health Organization |
| ART | Arterial pressure |
| WSN | Wireless sensor network |
| SMILE | Smart Independent Living for Elders |
| CSS | Coordinating Sink Station Unit |
| QoS | Quality of service |
| SAR | Specific absorption rate |
| FCC | Federal Communications Commission |
| ICNIRP | International Commission on Non-Ionizing Radiation Protection |
| TARA | Temperature Aware Routing Algorithm |
| LTR | Least Temperature Rise |
| ALTR | Adaptive Least Temperature Rise |
| LTRT | Least Total Route Temperature |
| CDR | Critical Data Routing |
| MAC | Media Access Control |
| SCMC | Sensor-Centric Monte-Carlo |
| TBFC | Tissue-Based Fixed Coordinator |
| 1HC | One-hop caching |
| EEG | Electroencephalogram |
| ECG | Electrocardiogram |
| EMG | Electromyography |

| | |
|---|---|
| BAN | Body area network |
| SPO2 | Peripheral capillary oxygen saturation |
| PID | Proportional-integral-derivative |
| ANN-NLR | Artificial neural network – non-linear regression |
| GPRS | General Packet Radio Service |
| WLAN | Wireless Local Area Network |
| TI | Texas Instruments |
| 1DSSS | Direct Sequence Spread Spectrum |
| 54OFDM | Orthogonal Frequency Division Multiplexing |

## Author details

Amitabh Mishra
University of West Florida, Pensacola, FL, USA

*Address all correspondence to: amishra@uwf.edu

## References

[1] United Nations. World Population Prospects: The 2010 Revision [Internet]. 2010. Available from: http://esa.un.org/wpp/ [Accessed: 2021-05-15]

[2] PricewaterhouseCoopers. Medical cost trend: Behind the numbers 2022 [Internet]. 2020. Available from: https://www.pwc.com/us/en/industries/health-industries/library/behind-the-numbers.html [Accessed: 2021-05-15]

[3] Peterson-KFF. How has U.S. spending on healthcare changed over time? [Internet]. 2021. Available from: https://www.healthsystemtracker.org/chart-collection/u-s-spending-healthcare-changed-time/#item-start [Accessed: 2021-05-15]

[4] Butler RN. Population aging and health. British Medical Journal. 1997; 315 (7115):1082–1084. DOI: 10.1136/bmj.315.7115.1082 [Accessed: 2021-05-15]

[5] Sartorius N. The meanings of health and its promotion. Croatian medical journal. 2006. 47(4): 662–664. PMCID: PMC2080455 [Accessed: 2021-05-15]

[6] Charakida M, Masi S, Deanfield JE. The year in Cardiology 2012: focus on cardiovascular disease prevention. European Heart Journal. 2013; 34(4): 314–317 Available from: https://doi.org/10.1093/eurheartj/ehs429 [Accessed: 2021-05-15]

[7] Berry JD, Dyer A, Cai X, Garside DB, Ning H, Thomas A, et al. Lifetime risks of cardiovascular disease. New England Journal of Medicine. 2012; 366(4):321–329 DOI: 10.1056/NEJMoa1012848 [Accessed: 2021-05-15]

[8] Kohl HW, Craig CL, Lambert EV, Inoue S, Alkandari JR, Leetongin G, et al. The pandemic of physical inactivity: global action for public health. Lancet. 2012; 380(9838):294–305 DOI: 10.1016/S0140-6736(12)60898-8 [Accessed: 2021-05-15]

[9] Gardner AW, Parker DE, Montgomery PS, Scott KJ, Blevins SM. Efficacy of quantified home-based exercise and supervised exercise in patients with intermittent claudication: a randomized controlled trial. Circulation 2011; 123(5):491–498 DOI: 10.1161/CIRCULATIONAHA.110.963066 [Accessed: 2021-05-15]

[10] Go AS, Hylek EM, Phillips KA, Chang Y, Henault LE, Selby JV, et al. Prevalence of diagnosed atrial fibrillation in adults: national implications for rhythm management and stroke prevention: the AnTicoagulation and Risk Factors in Atrial Fibrillation (ATRIA) study. Journal of the American Medical Association. 2001; 285(18):2370–2375 DOI: 10.1001/jama.285.18.2370 [Accessed: 2021-05-15]

[11] HuntSA, et al. ACC/AHA guidelines for the evaluation and management of chronic heart failure in the adult: executive summary a report of the American college of cardiology/American Heart Association task force on practice guidelines (committee to revise the 1995 guidelines for the evaluation and management of heart failure). J Am Coll Cardiol. 2001; 104 (24):2996–3007 DOI: 10.1016/s0735-1097(01)01683-7 [Accessed: 2021-05-15]

[12] Benjamin EJ, Wolf PA, D'Agostino RB, Silbershatz H, Kannel WB, Levy D. Impact of atrial fibrillation on the risk of death: the Framingham heart study. Circulation 1998; 98(10):946–952. DOI: 10.1161/01.cir.98.10.946 [Accessed: 2021-05-15]

[13] Shahoud JS, Sanvictores T, Aeddula NR. Physiology, Arterial Pressure Regulation. [Updated 2020 Sep

6]. StatPearls [Internet]. StatPearls Publishing. 2021. Available from: https://www.ncbi.nlm.nih.gov/books/NBK538509/ [Accessed: 2021-05-15]

[14] Carretero OA, Oparil S. Essential hypertension. Part I: definition and etiology. Circulation. 2000 Jan 25;101 (3):329-335 DOI: 10.1161/01.cir.101.3.329 [Accessed: 2021-05-15]

[15] Oparil S, Zaman MA, Calhoun DA. Pathogenesis of hypertension. Ann Intern Med. 2003 Nov 04;139(9): 761-776 DOI: 10.7326/0003-4819-139-9-200311040-00011 [Accessed: 2021-05-15]

[16] Chobanian AV, Bakris GL, Black HR, Cushman WC, Green LA, Izzo JLJ, et al. Seventh report of the joint national committee on prevention, detection, evaluation, and treatment of high blood pressure. Hypertension 2003; 42(6):1206–1252 DOI: 10.1161/01.HYP.0000107251.49515.c2 [Accessed: 2021-05-15]

[17] Rubel P, Fayn J, Simon-Chautemps L, Atoui H, Ohlsson M, Telisson D, et al. New paradigms in telemedicine: ambient intelligence, wearable, pervasive and personalized. Studies in Health Technology and Informatics. 2004; 108:123–132 PMID: 15718638 [Accessed: 2021-05-15]

[18] Chouvarda I, Koutkias V, Malousi A, Maglaveras N. Grid-enabled biosensor networks for pervasive healthcare. Studies in Health Technology and Informatics. 2005; 112: 90–99 PMID: 15923719 [Accessed: 2021-05-15]

[19] Jovanov E, Milenkovic A, Otto C, de Groen PC. A wireless body area network of intelligent motion sensors for computer assisted physical rehabilitation. Journal of NeuroEngineering and Rehabilitation. 2005; 2(6) DOI: https://doi.org/10.1186/1743-0003-2-6 [Accessed: 2021-05-15]

[20] Wang L, Johannessen EA, Hammond PA, Cui L, Reid SW, Cooper JM, et al. A programmable microsystem using system-on-chip for real-time biotelemetry. IEEE Transactions on Biomedical Engineering. 2005; 52(7):1251–1260 DOI: 10.1109/TBME.2005.847562 [Accessed: 2021-05-15]

[21] Tavel ME. Cardiac auscultation: a glorious past–and it does have a future!. Circulation 2006; 113(9):1255–1259 DOI: 10.1161/CIRCULATIONAHA.105.591149 [Accessed: 2021-05-15]

[22] Fragasso G, Cuko A, Spoladore R, Montano C, Palloshi A, Silipigni C, et al. Validation of remote cardiopulmonary examination in patients with heart failure with a videophone-based system. J Card Fail. 2007;13(4):281-286. DOI: 10.1016/j.cardfail.2007.01.008 [Accessed: 2021-05-15]

[23] Shantaram A, Beyenal H, Raajan R, Veluchamy A, Lewandowski Z. Wireless sensors powered by microbial fuel cells. Environmental Science and Technology. 2005; 39(13):5037–5042 DOI: 10.1021/es0480668 [Accessed: 2021-05-15]

[24] Bell G. Bell's law for the birth and death of computer classes. Communications of the ACM. 2008; 51 (1):86–94 DOI: https://doi.org/10.1145/1327452.1327453 [Accessed: 2021-05-15]

[25] Warneke B, Last M, Liebowitz B, Pister KSJ. Smart dust: communicating with a cubicmillimeter computer. Computer. 2001; 34(1):44–51 DOI: 10.1109/2.895117 [Accessed: 2021-05-15]

[26] Kahn JM, Katz RH, Pister KSJ. Next century challenges: mobile networking for smart dust. In: Proceedings of the International Conference on Mobile Computing and Networking (MobiCom '99); 15-19 August, 1999; Seattle. Boston: IEEE; 2000. P. 271–278 DOI:

https://doi.org/10.1145/313451.313558 [Accessed: 2021-05-15]

[27] Bulusu N, Jha S. Wireless sensor network systems: A systems perspective. illustrated ed. Artech House Publishers; 2005. 326 p. Available from: https://www.amazon.com/Wireless-Networks-Systems-Perspective-Sensors/dp/1580538673 [Accessed: 2021-05-15]

[28] Spencer BF, Jr, Cho S, Sim SH. Wireless monitoring of civil infrastructure comes of age. Structure Magazine. 2011;12–16. Available from: https://www.structuremag.org/?p=4531 [Accessed: 2021-05-15]

[29] HP. Shell and HP advance seismic sensing capabilities [Internet]. 2011. Available from: https://www.hp.com/us-en/hp-news/press-release.html?id=908010#.YN-U0OhKjRY [Accessed: 2021-05-15]

[30] Mainwaring A, Polastre J, Szewczyk R, Culler D, Anderson J. Wireless sensor networks for habitat monitoring. In: Proceedings of the First ACM International Workshop on Wireless Sensor Networks and Applications (WSNA '02); 28 September 2002; Atlanta. 2002. P. 88–97. DOI: https://doi.org/10.1145/570738.570751 [Accessed: 2021-05-15]

[31] Zhang S, McClean SI, Scotney BW. Probabilistic learning from incomplete data for recognition of activities of daily living in smart homes. IEEE Transactions on Information Technology in Biomedicine. 2012; 16(3): 454–462 DOI: 10.1109/TITB.2012.2188534. [Accessed: 2021-05-15]

[32] Chan M, Campo E, Esteve D, Fourniols J-Y. Smart homes – current features and future perspectives. Maturitas. 2009; 64:90–97 DOI: 10.1016/j.maturitas.2009.07.014 [Accessed: 2021-05-15]

[33] Reichherzer T, Mishra A, Kalaimannan E, Wilde N. A Case Study on the Trade-Offs between Security, Scalability, and Efficiency in Smart Home Sensor Networks. In: 2016 International Conference on Computational Science and Computational Intelligence (CSCI '16); 15-17 December 2016; Las Vegas. IEEE; 2016, p. 222-225 DOI: 10.1109/CSCI.2016.0049 [Accessed: 2021-05-15]

[34] Mishra A, Reichherzer T, Kalaimannan E, Wilde N, Ramirez R. Trade-offs involved in the choice of cloud service configurations when building secure, scalable, and efficient Internet-of-Things networks. International Journal of Distributed Sensor Networks. 2020; 16(2): 1-13. DOI: https://doi.org/10.1177/1550147720908199 [Accessed: 2021-05-15]

[35] Lo BPL, Yang GZ. Key technical challenges and current Implementations of body sensor networks. In: Proceedings of the Second International Workshop on Wearable and Implantable Body Sensor Networks (BSN 2005); April 2005; London. Washington, DC: IEEE; 2006. p. 1–5. Available from: https://www.doc.ic.ac.uk/ benlo/ubimon/BSN.pdf [Accessed: 2021-05-15]

[36] Hirata A, Shiozawa T. Correlation of maximum temperature increase and peak SAR in the human head due to handset antennas. IEEE T Microw Theory. 2003; 51:1834–1841 DOI: 10.1109/TMTT.2003.814314 [Accessed: 2021-05-15]

[37] Havenith G. Individualized model of human thremoregulation for the simulation of heat stress response. J. Appl. Physiol. 2001; 90:1943–1954. DOI: 10.1152/jappl.2001.90.5.1943 [Accessed: 2021-05-15]

[38] Natarajan A, de Silva B, Kok-Kiong Y, Motani M. To Hop or Not to Hop:

Network Architecture for Body Sensor Networks. In: Proceedings of the 6th Annual IEEE Communications Society Conference on Sensor, Mesh and Ad Hoc Communications and Networks (SECON); 22–26 June 2009; Rome. p. 1–9. Available from: https://www.ece.nus.edu.sg/stfpage/motani/pubs/Natarajan_Secon2009.pdf [Accessed: 2021-05-15]

[39] Lazzi G. Thermal effects of bioimplants. IEEE Eng. Med. Biol. 2005; 24:75–81. DOI: 10.1109/MEMB.2005.1511503 [Accessed: 2021-05-15]

[40] Punj R, Kumar R. Technological aspects of WBANs for health monitoring: a comprehensive review. Wireless Networks. 2019. 25(3):1125–1157 DOI: https://doi.org/10.1007/s11276-018-1694-3 [Accessed: 2021-05-15]

[41] Kathe KS, Deshpande UA. A Thermal Aware Routing Algorithm for a Wireless Body Area Network. Wireless Personal Communications. Springer. 2019; 105:1353–1380. DOI: https://doi.org/10.1007/s11277-019-06148-w [Accessed: 2021-05-15]

[42] Quwaider M, Biswas S. On-Body Packet Routing Algorithms for Body Sensor Networks. In: Proceedings of 1st IEEE International Conference on Networks & Communications (NetCoM 2009); 27-29 December 2009; Chennai. New York: IEEE; 2010. p. 171–177. DOI: 10.1109/NetCoM.2009.54 [Accessed: 2021-05-15]

[43] Quwaider M, Biswas S. DTN Routing in Body Sensor Networks with Dynamic Postural Partitioning. Ad Hoc Networks. 2010; 8(8): 824–841. DOI: https://doi.org/10.1016/j.adhoc.2010.03.002 [Accessed: 2021-05-15]

[44] Quwaider M, Biswas S. Probabilistic Routing in On-Body Sensor Networks with Postural Disconnections. In: Proceedings of 7th ACM International Symposium on Mobility Management and Wireless Access (MobiWac'09); 26-27 October, 2009; Tenerife, Canary Islands, Spain. 2009: p. 149–158. DOI: https://doi.org/10.1145/1641776.1641803 [Accessed: 2021-05-15]

[45] Tang Q, Tummala N, Gupta SKS, Schwiebert L. TARA: Thermal-Aware Routing Algorithm for Implanted Sensor Networks. In: Proceedings of 1st IEEE International Conference on Distributed Computing in Sensor Systems (DCOSS 2005); 30 June - 1July 2005; Marina del Rey, CA, USA. 2005: p. 206–217. Available from: https://citeseerx.ist.psu.edu/viewdoc/download?doi=10.1.1.68.4969&rep=rep1&type=pdf [Accessed: 2021-05-15]

[46] Maskooki A, Soh CB, Gunawan E, Low KS. Opportunistic Routing for Body Area Network. In: Proceedings of IEEE Consumer Communications and Networking Conference (CCNC '11); 9-12 January 2011; Las Vegas, NV, USA. 2011: p. 237–241. DOI:10.1109/CCNC.2011.5766463 [Accessed: 2021-05-15]

[47] Bag A, Bassiouni M. Energy Efficient Thermal Aware Routing Algorithms for Embedded Biomedical Sensor Networks. In Proceeding of IEEE International Conference on Mobile Ad Hoc and Sensor Systems (MASS '06); 9 - 12 October 2006; Vancouver, BC, Canada. 2006: p. 604–609. [Accessed: 2021-05-15]

[48] Takahashi D, Xiao Y, Hu F, Chen J, Sun Y. Temperature-Aware Routing for Telemedicine Applications in Embedded Biomedical Sensor Networks. EURASIP J. Wirel. Commun. Netw. 2008; 2008: p. 1–11. Available from: https://jwcn-eurasipjournals.springeropen.com/articles/10.1155/2008/572636 [Accessed: 2021-05-15]

[49] Djenouri D, Balasingham I. New QoS and Geographical Routing in Wireless Biomedical Sensor Networks.

In Proceedings of 6th IEEE International ICST Conference on Broadband Communications, Networks, and Systems (BROADNETS 2009); 14-16 September 2009; Madrid, Spain. 2009: p. 1–8. DOI: http://dx.doi.org/10.4108/ICST.BROADNETS2009.7188 [Accessed: 2021-05-15]

[50] Razzaque MA, Hong CS, Lee S. Data-Centric Multiobjective QoS-Aware Routing Protocol for Body Sensor Networks. Sensors (Basel). 2011; 11(1): 917–937. DOI: https://doi.org/10.3390/s110100917 [Accessed: 2021-05-15]

[51] Khan Z, Sivakumar S, Phillips W, Robertson B. QPRD: QoS-aware Peering Routing Protocol for Delay Sensitive Data in Hospital Body Area Network Communication. In: Proceedings of 7th IEEE International Conference on Broadband, Wireless Computing, Communication and Applications (BWCCA '12); 12-14 November 2012; Victoria, BC, Canada. 2012: p. 178–185. DOI: https://doi.org/10.1155/2015/153232 [Accessed: 2021-05-15]

[52] Khan ZA, Sivakumar S, Phillips W, Robertson B. A QoS-aware Routing Protocol for Reliability Sensitive Data in Hospital Body Area Networks. Procedia Comput. Sci. 2013; 19: 171–179. DOI: https://doi.org/10.1016/j.procs.2013.06.027 [Accessed: 2021-05-15]

[53] Monowar MM, Hassan MM, Bajaber F, Hamid MA, Alamri A. Thermal-Aware Multiconstrained Intrabody QoS Routing for Wireless Body Area Networks. Int. J. Distrib. Sens. Networks. 2014; 2014(676312): 1–14. DOI: https://doi.org/10.1155/2014/676312 [Accessed: 2021-05-15]

[54] Bangash JI, Abdullah AH, Razzaque MA, Khan AW. Reliability Aware Routing for Intra-Wireless Body Sensor Networks. Int. J. Distrib. Sens. Networks. 2014; 2014(786537): 1–10. DOI: https://doi.org/10.1155/2014/786537 [Accessed: 2021-05-15]

[55] Bangash JI, Abdullah AH, Khan AW, Razzaque MA, Yusof R. Critical Data Routing (CDR) for Intra Wireless Body Sensor Networks. TELKOMNIKA. 2015. 13(1):181-192. DOI: 10.12928/TELKOMNIKA.v13i1.365 [Accessed: 2021-05-15]

[56] Krishnamurthy AG, Jun JH, Agrawal DP. Gradient based Temperature-aware routing in Body Area Sensor Networks. 9th International Conference on Body Area Networks (BodyNets '14); 29 September – 1 October 2014; London. ICST: 2014. p. 217-221. DOI: http://dx.doi.org/10.4108/icst.bodynets.2014.257002 [Accessed: 2021-05-15]

[57] Cashwell ED, Everett CJ. A practical manual on the Monte Carlo Method for random walk problems. New York, 1960. Pergamon Press. 40(7-8): 379-380. DOI: https://doi.org/10.1002/zamm.19600400727 [Accessed: 2021-05-15]

[58] Silay KM, Dehollaini C, Declercq M. Numerical analysis of temperature elevation in the head due to power dissipation in a cortical implant. 30th Annual International Conference of the IEEE Engineering in Medicine and Biology Society (EMBS 2008); 20-25 August 2008; Vancouver, BC, Canada. IEEE; 2008. p.951-956. DOI: 10.1109/IEMBS.2008.4649312 [Accessed: 2021-05-15]

[59] Mishra A, Agrawal DP. Evaluation of Suitability of Current Industrial Standards in Designing Control Applications for Internet of Things Healthcare Sensor Networks. Journal of Sensor and Actuator Networks. MDPI. 2019. 8(4); 54; p. 1-20. DOI:10.3390/jsan8040054 [Accessed: 2021-05-15]

[60] Physionet. The Research Resource for Complex Physiologic Signals [Internet]. 1999. Available from: http://www.physionet.org [Accessed: 2021-05-15]

[61] Mercy AC, Renold PA. Routing protocol for low power lossy networks. 2014 IEEE International Conference on Advanced Communications, Control and Computing Technologies (ICACCCT '14); 08–10 May 2014; Ramanathapuram, Tamil Nadu, India. 2014. pp. 1457-1461. DOI: 10.1109/ICACCCT.2014.7019343 [Accessed: 2021-05-15]

[62] Pennes HH. Analysis of Tissue and Arterial Blood Temperatures in the resting human forearm. Journal of Applied Physiology. 1948; 1: 93-121, 1948. DOI: https://doi.org/10.1152/jappl.1948.1.2.93 [Accessed: 2021-05-15]

[63] Takahashi D, Xiao Y, Hu F. LTRT: Least Total-Route Temperature routing for Embedded Biomedical Sensor Networks. In: IEEE Global Telecommunications Conference, 2007 (GLOBECOM '07); 28-30 November 2007; Washington, D.C. IEEE; 2007. p. 641–645. DOI:10.1109/GLOCOM.2007.125 [Accessed: 2021-05-15]

[64] Tabandeh M, Jahed M, Ahourai F, Moradi S. A thermal-aware shortest hop routing algorithm for in vivo biomedical sensor networks. In: Proceedings of the 2009 Sixth International Conference on Information Technology (ITNG '09); 27-29 April 2009; Las Vegas, Nevada, USA. IEEE; 2009. p. 1612–1613. DOI: https://doi.org/10.1109/ITNG.2009.274 [Accessed: 2021-05-15]

[65] Bouldjadj S, Aliouat Z. High Throughput and Thermal Aware Routing Protocol (HTTRP) for Wireless Body Area Networks. Ingénierie des Systèmes D Information. 2020; 25(1): 41-48. DOI: 10.18280/isi.250106 [Accessed: 2021-05-15]

[66] Ullah S, Higgin H, Siddiqui MA, Kwak KS. A study of implanted and wearable body sensor networks. In: 2nd KES International Conference on Agent and Multi-Agent Systems: Technologies and Applications (KES-AMSTA'08); 26-28 March 2008; Incheon, South Korea. Springer-Verlag, Berlin Heidelberg; 2008. p. 464-473. DOI: https://doi.org/10.1007/978-3-540-78582-8_47 [Accessed: 2021-05-15]

[67] Dishman E. Inventing wellness systems for aging in place. Computer. 2004; 37(5): 34-41. DOI: 10.1109/MC.2004.1297237 [Accessed: 2021-05-15]

[68] Hanson MA, Powell HC, Barth AT, Ringgenberg K, Calhoun BH, Aylor JH, Lach J. Body area sensor networks: Challenges and opportunities. Computer. 2009. 42(1): 58-65. Available from: https://www3.nd.edu/ dwang5/courses/spring19/papers/bsn/BSNOverview.pdf [Accessed: 2021-05-15]

[69] Mishra A, Chakraborty S, Li H, Agrawal DP. Error Minimization and Energy Conservation by predicting data in Wireless Body Sensor Networks using Artificial Neural Network and Analysis of Error. In: 2014 IEEE 11th Consumer Communications and Networking Conference (CCNC-2014); 10-13 January 2014. Las Vegas, NV, USA. IEEE: 2014. P.164-170 DOI: 10.1109/CCNC.2014.7056324 [Accessed: 2021-05-15]

[70] Mishra A, Agrawal DP. Energy Conservation and Lifetime Optimization of Wireless Body Sensor Networks for 24x7 Physiological parameters' Monitoring. Journal of Communications. 2015; 10(9):685-695. DOI: 10.12720/jcm.10.9.685-695. [Accessed: 2021-05-15]

[71] The MathWorks Inc. MATLAB, version (R2020). Natick, Massachusetts: 2020. https://www.mathworks.com/ [Accessed: 2021-05-15]

[72] Tie SF, Tan CW. A review of energy sources and energy management system in electric vehicles. Renewable and

Sustainable Energy Reviews. Elsevier. 2013; 20(C):82–102. DOI: 10.1016/j.rser.2012.11.077 [Accessed: 2021-05-15]

[73] Holladay JD, Jones EO, Phelps M, Hu J. Microfuel processor for use in a miniature power supply. J. Power Sources. Elsevier. 2002; 108(1–2): Pages 21–27. DOI: https://doi.org/10.1016/S0378-7753(01)01011-4 [Accessed: 2021-05-15]

[74] Kim T, Kwon S. MEMS fuel cell system integrated with a methanol reformer for a portable power source. Sensors and Actuators A: Physical. Elsevier. 2009; 154(2):204–211. DOI: 10.1016/j.sna.2008.07.010 [Accessed: 2021-05-15]

[75] Cho J-H, Lin CS, Richards CD, Richards RF, Ahn J, Ronney PD. Demonstration of an external combustion micro-heat engine. Proceedings of PowerMEMS 2008+ microEMS2008; 9-12 November 2008; Sendai, Japan. Proc. Combustion Institute. 2009; 33(2): 3099–3105. DOI: 10.1016/J.PROCI.2008.07.017 [Accessed: 2021-05-15]

[76] Wee JH. A feasibility study on direct methanol fuel cells for laptop computers based on a cost comparison with lithium-ion batteries. J. Power Sources. 2007; 173:424–443. DOI:10.1016/J.JPOWSOUR.2007.04.084 [Accessed: 2021-05-15]

[77] Kerzenmacher S, Kra¨ling U, Metz T, Zengerle R, von Stetten F. A potentially implantable glucose fuel cell with Raney-platinum film electrodes for improved hydrolytic and oxidative stability. J. Power Sources. 2011; 196 (3):1264–1272. DOI: 10.1016/j.jpowsour.2010.08.019 [Accessed: 2021-05-15]

[78] Heinzelman W. Application specific Protocol Architectures for Wireless Networks [thesis]. Boston: Massachusetts Institute of Technology; 2010. [Accessed: 2021-05-15]

[79] Park C, Liu J, Chou PH. ECO: an ultra-compact low-power wireless sensor node for real-time motion monitoring. Fourth International Symposium on Information Processing in Sensor Networks 2005 (IPSN 2005); 24-27 April 2005; Los Angeles California. IEEE; 2005. p. 398-403. DOI: 10.1109/IPSN.2005.1440956 [Accessed: 2021-05-15model has the below assumptions]

[80] TI. CC3100 SimpleLink Wi-Fi Network Processor, Internet-of-Things Solution for MCU Applications (Rev. D) [Internet]. 2014. Available from: http://www.ti.com/product/CC3100/datasheet [Accessed: 2021-05-15]

Chapter 6

# Environmental Management and Data for the SDGs

*Luiz Cruz Villares*


## Abstract

Ambient intelligence data is key for the global scope of climate, biodiversity, water, air pollution, and general environmental control. The account of carbon emissions, water and air pollutants, sea and freshwater life, and earth nature quality, are key for a healthy environment, in local or global concerns. Controlling initiatives and data management for the environment are part of sustainability strategies for governments, businesses, and entities in general. Such initiatives are increasingly connected to key mankind issues such as poverty, hunger, health, education, gender and equality, among others. Altogether, they are clearly addressed in the United Nations Sustainable Development Goals, to be met in 2030. The Sustainable Development Goals (SDGs) provide a relevant connection of environmental intelligence systems with public policies, research, international cooperation, and technology transfers. In a detailed analysis of their targets, they offer the stimulus for a vast array of technical applications linking sustainability issues with Ambient Intelligence technologies. The SDGs should be considered the benchmark for action towards a healthy environment given their global concern and interconnectivity of multiple issues related to a better quality of life. The SDGs should become, ultimately, the main driver for the spread of present and new technologies to promote better monitoring and control of the environment. The growth and advances of Ambient Intelligence technologies to new areas of knowledge should be enhanced by their connectivity with the SDGs.


**Keywords:** Sustainable Development Goals, Ambient Intelligence, AmI, targets, connections, interdependence, international cooperation, technology

## 1. Introduction

This chapter identifies several targets clearly outlined in the SDGs to be implemented with straight cooperation with Ambient Intelligence technologies. Such relationships shall promote a high level of Ambient Intelligence research, and the growth of technical and international cooperation within the Ambient Intelligence community for a better and more sustainable planet.

Following this introduction, this chapter has two parts and a conclusion, in brief, covering a description of the SDGs; a detailed analysis of each SDGs' targets and indicators linking with Ambient Intelligence technologies, and a conclusion.

The first part offers a primer on the SDGs, to set a common ground of knowledge about their importance and reach of all sustainability issues worldwide. Part two is the body of study in this chapter, presenting analyzes of all links among the SDGs and Ambient Intelligence issues. Each SDG is presented with a selection of targets presenting connections with Ambient Intelligence, in technology, public policy, and general cooperation aspects. Given its general awareness, SDG 13, related to climate change receives a comprehensive attention. Part three is a conclusion of this study, aimed at raising the general awareness of research, development, and cooperation efforts, worldwide, on Ambient Intelligence issues for the SDGs to be successfully achieved by 2030.

## 2. Key aspects of the SDGs for ambient intelligence

The Sustainable Development Goals (SDGs) are the result of a United Nations Organization (UN) initiative for all its Member States to reach global sustainable development standards in 2030 [1]. Created in 2015 as a development from the Millennium Development Objectives, the SDGs are composed of 17 major goals with 169 targets and specific indicators for measuring their achievements, by 2030. The SDGs represent a clear guide to all humanity to reach high sustainable development patterns in this present decade. The implementation of these targets are addressed at national and subnational government policies, and non-government initiatives. Businesses in general, mainly large corporations, have been adopting the SDGs in their sustainability strategies, committed to the most applicable targets related to their operations, activities and socio environmental impacts. In brief, the global 17 goals for 2030 are (**Figure 1**):

1. End poverty in all its forms, everywhere.

2. End hunger, achieve food security and improved nutrition, and promote sustainable agriculture.

3. Ensure healthy lives and promote well-being for all at all ages.

4. Ensure inclusive and equitable quality education and promote lifelong learning opportunities for all.

5. Achieve gender equality and empower all women and girls.

6. Ensure availability and sustainable management of water and sanitation for all.

7. Ensure access to affordable, reliable, sustainable and modern energy for all.

8. Promote sustained, inclusive and sustainable economic growth, full and productive, employment and decent work for all.

9. Build resilient infrastructure, promote inclusive and sustainable industrialization, and foster innovation.

10. Reduce inequality within and among countries.

11. Make cities and human settlements inclusive, safe, resilient and sustainable.

12. Ensure sustainable production and consumption patterns.

13. Take urgent action to combat climate change and its impacts.

14. Conserve and sustainably use the oceans, seas and marine resources for sustainable development.

15. Protect, restore and promote sustainable use of terrestrial ecosystems, sustainably manage forests, combat desertification, and halt and reverse land degradation, and halt biodiversity loss.

16. Promote peaceful and inclusive societies for sustainable development, provide access to justice for all and build effective, accountable and inclusive institutions at all levels.

17. Strengthen the means of implementation and revitalize the global partnership for sustainable development.

**Figure 1.**
*UN sustainable development goals (SDGs). Source: United Nations.*

As mentioned, each goal has specific targets with specific actions, totaling 169 targets. Overall, the targets are interconnected and complementary to each other, pointing to similarities with multiple cross-references among them, thus creating synergies for their implementation. The targets are enhanced by actions with common terms such as "reduce", "increase", "assure", "promote", "implement", "ensure" "halt", "improve", "support", "end", improve", among others [2].

These actions create interdependence among them, pointing to public policies, scientific research, funding and international cooperation issues. In this chapter, attention is placed on the SDGs' focus on environmental targets, hence closer to Ambient Intelligence. The standard analysis provided is the identification of targets connectable to Ambient Intelligence technologies as a general overview. In some cases, specific mentions are made to technologies, but the overall proposal in this study is to enlighten the SDGs and targets as major drivers to develop and spread the Ambient Intelligence solutions for sustainable development (**Figure 1**).

Following, this chapter is focused on listing each SDG and its specific goals related to Ambient Intelligence technologies. In some cases, a SDG may be not strictly related to the environment, but it brings specific targets related to Ambient Intelligence subjects. Henceforth, in the next part, Ambient Intelligence is abbreviated to "AmI".

## 3. Connections of the SDGs targets with ambient intelligence

Beginning, SDG 1, related to poverty, indicates general targets for humanity to reduce poverty, proposing overall measures and tasks, generally interconnected with all SDGs. As an example, Target 1.5 is about improving the resilience of the poor and all population in vulnerable situations, and reduce their exposure and vulnerability to climate-related extreme events, as well as other forms of economic, social and environmental disasters. This statement brings connections with most of the other 16 SDGs, and specifically, calls for the improvement of general means of living involving the internet, connectivity, and the use of collective solutions for housing solutions to be fairly improved with AmI devices, such as smart energy consumption devices.

SDG 2 is related to zero hunger. Target 2.4 is to increase the production of food from sustainable and resilient agricultural practices, helping the maintenance of ecosystems, and the strength of human capacity to face extreme weather situations, such as drought, flooding and other disasters, thus improving land and soil quality. The indicator to measure achievements by 2030 is the proportion of agricultural areas under productive and sustainable agriculture. The main target here is to improve the proportion of sustainable land practices over traditional land practices. AmI devices provide multiple sensors for climate and land systems controls among other monitoring tools to enhance the natural resource management attached to good practices of land management. Weather predictions coupled with crop productivity and biodynamic techniques use AmI technologies for productivity, water and natural resources uses. Next, Target 2.5 calls for an immediate action, already by 2020, on the maintenance of the genetic diversity of seeds, cultivated plants, domesticated animals and their related wild species. Continuing, the target calls for the implementation of seed and plant banks at national, regional and international levels, and the management of equitable access to genetic resources associated to traditional knowledge. The indicators are the number of plant and animal genetic resources for food and agriculture secured in either medium or long-term conservation facilities; and the proportion of local breeds classified as being at risk, not-at-risk or at unknown level of risk of extinction. These are ultimate areas for AmI devices to help a proper registry of species in the form of local and global banks of species, mainly, using Blockchain platforms to ensure data security and veracity. Biodiversity conservation and food security issues have increasingly become part of integrated efforts. These specifically mentioned SDG 2 targets are closely related to SDG 15, about life on earth.

SDG 3, related to good health and well-being, apparently provides connectivity to AmI in areas of public health, presently enhanced by the global Covid 19 pandemic, bringing much attention to AmI possibilities to cooperate with safer and cleaner ambients. For instance, cognitive technologies, such as machine learning, neural networks, robotic process automation, bots, neural nets, and the broader domain of AmI, have the potential to transform the human relation with critical health and safety situations, presently enhanced by all attention to Covid-19. In brief, AmI offers tools for medical professionals to have accurate predictions, fast reactions and interactions with critical health and safety situations [3].

An examination of specific targets reinforces the connections between AmI, health and sustainable development. Target 3.6 calls, already by 2020, for a decrease in the number of global deaths and injuries from road traffic accidents. The indicator is the death rate due to road traffic injuries. AmI and Internet of Things (IoT) devices installed in cars, in some cases connectable to road management, shall decrease the number of road accidents in the coming years, thus positively contributing to this target. Continuing, Target 3.9 has clear connections with AmI, as it aims at substantially reducing deaths and illnesses from hazardous chemicals and air, water and soil pollution and contamination. Water pollutants, hazardous chemicals and metals, and air pollution measurements, are clear objects of AmI devices and systems, already playing a major role in their monitoring. Air pollution, both ambient and household, increases the risk of cardiovascular and respiratory disease with above seven million deaths worldwide. For instance, at Sub-Saharan Africa and most of Asia and Oceania (excluding Australia/New Zealand) high mortality rates are caused by air pollution, where a meaningful proportion of the population employ polluting fuels and technologies for cooking. Inadequate health situations are generally linked to unsafe drinking water, sanitation and hygiene. For instance, 60 percent of the disease occurrences from diarrhea, 100 percent of infections from soil-transmitted helminths, and 16 percent of the burden from protein-energy malnutrition. Overall statistics point to a total of 870,000 deaths in 2016, from the above conditions [4]. These alarming figures point to strong AmI efforts to minimize the occurrence of these diseases, with technology and international cooperation issues.

SDGs 4 and 5 are related to quality education and gender equality, representing areas of key importance for sustainable development progress. The connection of AI devices with education and gender equality issues shall be indirectly related to the environment. The scope of this chapter is focused on primary connections among AmI technologies and the SDGs. Therefore, no further mention about their targets is made in the scope of this study.

SDG 6 relates to clean water and sanitation. It presents a complement to SDG 3, in particular, on water pollution and contaminants, expressing the interconnection among the SDGs. Most of its targets call for clean water and sanitation for the entire global population. AmI devices play a fundamental role in monitoring water quality and wastewater sanitation. For example, Target 6.3 is about reducing the water pollution, the elimination of dumping and minimization of hazardous chemicals releases, avoiding the increase of untreated wastewater at the same time the water recycling and reuse should be globally increased. The indicators for achieving this target are: proportion of wastewater safely treated and proportion of bodies of water with good ambient water quality. This target is probably among the most connected to AmI in the entire SDGs group of targets. Next, Target 6.4 is about water-use and fresh water supply efficiency aiming at reducing the number of people suffering from water scarcity. The indicators are change in water-use efficiency over time and the level of water stress: freshwater withdrawal as a proportion of available freshwater resources. These are technical issues related to water usage efficiency, with relevant technology from AmI devices and data systems.

SDG 7 is related to affordable and clean energy. Targets 7.1, 7.2 and 7.3 mention, by 2030, the availability of universal access to affordable, reliable and modern energy services, the growth of renewable energy share in the total final energy consumption, and doubling the global rate of improvement in energy efficiency. It is fairly possible to envision a positive contribution of AmI systems related to home improvement, with energy savings measures from smart consumption practices, increasingly from non-fossil sources, such as solar energy using retrofit and collective exchanges (smart collective grids). These technologies may increasingly use Blockchain platforms to ensure safe consumption registers among all users (affordable energy). Smart grids, combined with energy sources and exchanges, such as solar power generation controls and other devices are becoming more present in modern buildings and houses. The Covid 19 pandemic has boomed the adoption of home office work, hence the need for further attention to energy management and savings. Additionally, the rise of cryptocurrencies has provided the tokenization and other non-fiduciary ways of monetizing energy trades and a safe accounting for the exchanges and information controls among users [5].

SDG 9, related to industry, innovation and infrastructure presents many links to AmI. For instance, target 9a is about the facilitation of sustainable and resilient infrastructure development in developing countries through enhanced financial, technological and technical support to African countries, least developed countries, landlocked developing countries, and small island developing States. Target 9b is about the support of domestic technology development, research and innovation in developing countries, including the enforcement of a conducive environment policy for industrial diversification and value addition to commodities. The indicators are about the proportion of medium and high-tech industry value added in total value added. Target 9c aims at significantly increasing the access to information and communications technology, and strive to provide universal and affordable access to the Internet in least developed countries by 2020. The indicators are about the proportion of population covered by a mobile network and related technology. Notice the target date to be 2020, hence, already late in this achievement.

Additionally, target 9.4 calls for an upgrade and retrofit of industries with better resources' efficiency, improved technologies and processes, with more environmentally and clean solutions, according to each country capability level. The indicators are about CO2 emission per unit of value added. Next, Target 9.5 is dedicated to scientific research, technological upgrades of industrial sectors, particularly in developing countries, with the increase of personnel dedicated to research and development; and the growth of private and public research spending. The indicators are about research and development expenditure as a proportion of GDP.

SDG 10 relates to the reduction of inequality among and within countries. From an AmI perspective, it presents possible indirect connections to technologies for social and economic growth. Albeit it brings central issues to mankind, within the scope of this chapter, they are only indirectly related to AmI.

SDG 11, related to sustainable cities and communities, presents relevant interdependence with other SDGs and related AmI technologies. For example, Target 11.1 is about the access to adequate, safe and affordable housing for all, and the universalization of basic housing services and the upgrade of slums. This target is clearly connected to SDG 1 targets related to the availability and safety of housing, where

AmI plays a role in energy efficiency. Following, Targets 11.2 and 11.5 calls for a meaningful reduction of deaths and injuries from road disasters, water-related, and other disasters, with a focus on protecting the poor and people in vulnerable situations. The indicators are about the number of deaths, missing persons and persons affected by disasters and direct disaster economic loss in relation to global GDP. The indicator also includes the account of disasters′ damage to critical infrastructure and disruption of basic services. Target 11.6 is about the reduction of environmental impacts in cities, with special attention to air quality and municipal waste management. This Target is directly linked to AmI technologies. The indicators are clear about the ambient improvement, likely stated in the proportion of urban solid waste regularly collected and with adequate final discharge out of total urban solid waste generated, by cities; and annual mean levels of fine particulate matter (e.g. PM 2.5 and PM10) in cities (population weighted).

SDG 11 also presents other targets related to public policies about green areas and sustainable cities, indirectly benefited by AmI technologies. These targets also mention the areas air pollution and waste water, part of SDGs 3 and 6, being central for AmI technologies. Lastly, a mention is made to Target 11c, already due by 2020, that calls for a substantial increase in the number of cities and human settlements adopting and implementing integrated policies and plans towards inclusion, resource efficiency, mitigation and adaptation to climate change, resilience to disasters. The indicators are related to public improvements in disaster preparedness, increasingly observed in the recent years of climate change observation, thus connected to SDG 13.

SDG 12, related to responsible production and consumption, presents relevant issues related to AmI. Target 12.2 states overall ambitions for a clean environment, as it calls for the achievement of sustainable management and efficient use of natural resources, in this case, related to clean production, with indicators enhancing local production and consumption. Next, Target 12.3 calls for the end of global food waste at the retail and consumer levels, and the reduction of food losses along production and supply chains, including post-harvest losses, with indicators of food loss index. In such cases, sustainable agriculture, industry, and retail practices are challenged to improve their food efficiency, being AmI monitoring and data management part of the solutions for this implementation. This target clearly relates to SDG 2 about food security, sustainable agriculture, hence being common targets between these two SDGs. Following, Target 12.4 calls, already by 2020, for the achievement of an environmentally sound management of chemicals and all wastes throughout their life cycle. These practices should be in accordance with agreed international frameworks, and have a significant reduction of their release to air, water and soil, in order to minimize their adverse impacts on human health and the environment. The indicators clearly point to AmI solutions, as an achievement of improved multilateral environmental agreements on hazardous waste, and other chemicals, with commitments and obligations in transmitting information as required by each relevant agreement. An additional indicator is about the volume of hazardous waste generated per capita and proportion of treated hazardous waste volume. Target 12.5 adds to the former, calling for a substantial reduction of waste generation through prevention, reduction, recycling and reuse, with indicators about the volume recycled materials. Clearly, good waste management practices are supported by AmI technologies.

The remaining SDG 12 targets are more related to public policies. A special mention, however, is made about targets 12.7 and 12.8 that call for the promotion of public procurement practices that are sustainable, in accordance with national policies and priorities, and ensure that people everywhere have the relevant information and awareness for sustainable development and lifestyles in harmony with nature. The indicators are about the number of countries implementing sustainable public procurement policies and action plans. A special mention is made for this target, because it suggests an overall effort of all nations to implement clean production and consumption policies and extension to global citizenship education, including the education for sustainable development with emphasis on climate change. As mentioned before, within the SDGs goals, many targets are related to public policies and international cooperation. These specific targets may represent the most far reaching concerns for sustainable development practices to be adopted. In this sense, research and development programs and incentives are a key part for SDG 12 to be successful by 2030. AmI development must be clearly part of this effort.

SDG 13 relates to climate action, a key goal for the entire global sustainable development. Climate change is widely perceived as the greatest threat to sustainable development of the 21st century. The monitoring and mitigation of GreenHouse Gases[1] (GHGs) emissions have multiple links with AmI, in areas of weather data predictions, disasters' avoidance, as well as information tools for positive climate actions, regarding the accounting of carbon footprints, carbon accounts and trading, and other GHGs accounting.

The international cooperation efforts for climate action has more than 25 years of activities, led by the UN Conferences of the Parties (COPs). As openly noticed, the diplomatic efforts among nations to negotiate world climate agreements have not been successful in providing the necessary progress to stop growing annual temperatures, substantially correlated to the increase in particulate matters per million (PPMs), with observed consequences of melting ice caps, rising sea levels, among many others.

SDG 13 is aimed at bringing more efforts on the global concern to stop climate change in the Earth. The adaptation to weather impacts by countries and economic sectors, and the mitigation measures required to reduce and stop global warming are the key issues in the climate change agenda. The transition of fossil based energies to low-carbon economies presents multiple implications for economic and social development, production and consumption patterns, and labor, in general.

SDG 13, beyond its sole importance, presents a high interdependence and connectivity with other SDGs. The targets call for integrated global and regional actions, bringing most of the other SDGs to the solutions.

Target 13.1 is about enhancing human resilience to face the growing number of natural disasters, certainly related to SDG 11, about sustainable cities, among other goals. Indicators are about the number of countries adopting natural disasters risk reduction strategies, and others. Such efforts shall have the input of AmI for weather and atmosphere monitoring. For instance, researches show that birds may be able to anticipate the severity of the hurricane season ahead, months in advance [6]. Relevant studies about the timing of bird migration habits in anticipation of the hurricanes season in North America shows relevant connections between animal species intelligence and the right time of a storm approach. Targets 13.2 and 13.3 mention the integration of climate change measures into national policies, strategies and planning, and the improvement of education, awareness-raising, human, and institutional capacity on climate change mitigation, adaptation, impact reduction and early warning on the forecast impacts of climate change.

AmI, coupled with IoT and Artificial Intelligence present relevant issues in carbon measurement and pollution. Technologies related to machine learning can be a powerful tool in reducing GHG emissions. The areas of energy smart grids

---

[1] The primary greenhouse gases in Earth's atmosphere are water vapor (H2O), carbon dioxide (CO2), methane (CH4), nitrous oxide (N2O), and ozone (O3)

present many controls related do AmI, with connections to energy savings and redistribution. Disaster management systems, through AmI devices, can identify extreme weather impact problems with collaboration of machine learning features, in collaboration with other fields. Further research is fully recommended in these areas, which, as well, presents promising non fossil business opportunities [7].

The following targets in SDG 13 call for finance and international cooperation for humanity to face global warming. A special mention is made to forest and agriculture GHG emissions and management, being such subjects relevant to other SDGs, such as SDG 2, for sustainable agriculture practices, and SDG 15, for natural resources management. Presently, most economic policies do not place attention on clean and low carbon rural production practices. Most of the prevailing agriculture practices have been releasing more carbon to the atmosphere than in the past decades, with new crop areas coming from deforestation and unsustainable agricultural practices. On top of these effects, cattle and rice farming generate methane, a GHG with high Global Warming Potential (GWP). Overall, land use by humans is estimated, by many sources, to be responsible for about a quarter of global GHG emissions. In addition, it is widely observable the effect of permafrost melt, peat bogs dryness, and the increase of forest fires, becoming more frequent as a consequence of climate change itself, thus, all of which, releasing more carbon to the atmosphere [8]. The large scale of this problem allows for a similar scale of positive impact. Reductions could come from better land management and agriculture. AmI can help data monitoring in areas of precision agriculture to reduce carbon and methane releases from the soil and livestock, improving crop yields with low GHG emissions, without the need for deforestation. Satellite images make it possible to estimate the amount of carbon sequestered in a given area of land, as well as track GHG emissions from it. Machine learning, coupled with AmI, can help the monitoring of standing forests and peatlands, and predict the risk of fires, hence contributing to sustainable forestry practices [9]. These are also applicable issues to SDG 15, enhancing the interdependence of SDG 13 with all other SDGs.

Other relevant climate change issues taking place worldwide are the concerted implementation of carbon trading platforms allowing a safe and true register of carbon emissions and trade among citizens, business and governments, in line with a commodity trade market. The Paris Agreement at COP 21 in December 2015, set the path to promising efforts to climate change action, mainly in allowing the development of carbon markets based on reduction and carbon neutralization mechanisms, requiring the development of market and accounting standards, supported by trustable electronic bases. The technology under these trading platforms are under development using Blockchain based networks to assure a safe register of any carbon asset to be traded and stored, being it a financial asset, or only a right of ownership. This framework of programs and electronic rules will receive inputs from Spatial Web and Web 3.0 programming [10], providing possible connections to AmI controls for further assurances. These actions are taking place through international voluntary cooperation among programmers, traders, systems developers, and carbon trading professionals. The expected results are the creation of a reliable platform to account and record all voluntary carbon trading transactions, worldwide [11]. The voluntary carbon markets are expected to double in size in the upcoming years, with an exponential growth in ten years. AmI technologies shall be a relevant contributor to the checks and balances within these sophisticated and open systems to come.

SDG 14, related to life below water, presents good connections with AmI, such is the case of Target 14.1, about prevention and reduction of all kinds of marine pollution, in particular from land-based activities, including marine debris and nutrient pollution, with indicators about an index of coastal eutrophication and floating plastic debris density. Target 14.3 calls for the minimization of ocean acidification, calling nations to address such impact, including scientific cooperation at all levels. Target 14.4 refers, already by 2020, to an effective regulation of overfishing, with issues on illegal, unreported and unregulated fishing, and destructive fishing practices. The target also mentions the urgent implementation of science-based management plans to restore fish stocks, respecting the levels of maximum sustainable yield per species, as determined by their biological characteristics. Target 14.a is aimed at increasing scientific knowledge, developing research capacity, and transferring marine technology among nations. These actions must take into account the Intergovernmental Oceanographic Commission Criteria and Guidelines on the Transfer of Marine Technology. Such issues are aligned to improve ocean health and to enhance the contribution of marine biodiversity, for nations to improve their sea management and fishing practices, with special attention to small island developing States and least developed countries. The other targets in SDG 14 are about the regulations, funding and international cooperation for sustainable oceans and responsible fisheries. In all technical aspects within ocean acidification and fishing management, AmI must add relevant information for water and acidification, and eutrophication measures, among other issues.

Finally, SDG 14 has close connections with SDG 6, about water and sanitation, and SDG 13, related to climate change, once more, enhancing the interdependence among sustainable development issues (**Figure 2**).

SDG 15, related to life on land, is of major significance for the global environment and climate, as well as for populations and the economy. It presents a high interdependence among SDGs related to hunger, health, education, water, economy, inequalities, sustainable cities, climate change and life on water. The world

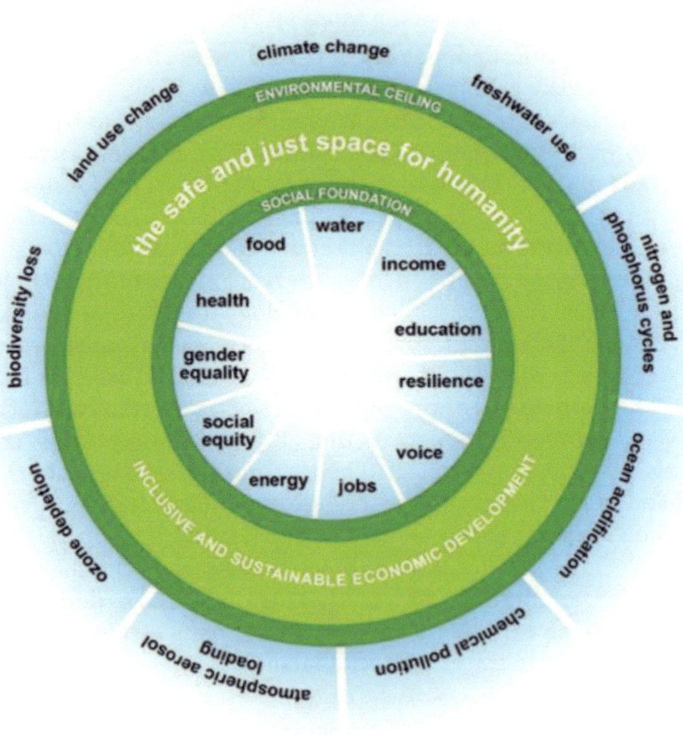


**Figure 2**
*Sustainable development interdependence. Source: https://www.fdsd.org/the-challenge/what-is-sustainable-development/.*

depends on standing forests and the other biomes with high ecological value to support most of the other SDGs. Target 15.1 calls, already in 2020, for the conservation, restoration and sustainable use of terrestrial and inland freshwater ecosystems and their services, in particular forests, wetlands, mountains and drylands. Since this target should be already in place, it means that the world should not have more deforestation at present times. The reality however, is that between 2015 and 2020, the rate of global deforestation was estimated at 10 million hectares per year, down from 16 million hectares per year in the 1990s. The area of primary forest coverage worldwide has decreased by over 80 million hectares since 1990. Deforestation and forest degradation continue to take place at alarming rates, which contributes significantly to the ongoing loss of biodiversity [12].

AmI must play an important role in imagery and biodiversity monitoring. The indicators are clear: forest area as a proportion of total land area, and proportion of important sites for terrestrial and freshwater biodiversity that are covered by protected areas, by ecosystem type. Target 15.2 complements the former, where it calls, also by 2020, for the implementation of sustainable management of all types of forests, the end of deforestation, the restoration of degraded forests, and for the substantial global growth of afforestation and reforestation. The indicators point to a progress in forest management. Such is the case for AmI technologies to support standing forests, mainly in the imagery, disaster preparedness and biodiversity issues.

The remaining targets in SDG 15 are related to desert, rivers and mountains conservation and restoration. Also, they call for equitative sharing of benefits arising from the utilization of genetic resources, actions against illegal trade and traffic of natural species. Target 15.8, for instance, calls for the introduction of measures to prevent the introduction and significantly reduce the impact of invasive alien species on land and water ecosystems and control or eradicate the priority species. The monitoring and safe register of genetic resources and control of invasive species presents important challenges to be met with AmI technologies, for example, with the use of banks of codes, supported by safe registers under Blockchain platforms and environmental databases showing constant update and progress analyzes. Finally, the areas of sustainable soil management, forest conservation, ecosystem services and all natural resources offer multiple uses for data, image, and monitoring management, based on AmI.

SDG 16 is related to peace, justice and strong institutions, with issues related to human rights, such as the assurance of citizenship, equality; to end corruption, end violence and promote equality among all. It is closely related to SDGs 1, 5 and 10, for instance, with focus on social issues. For this chapter overview of AmI relationships with the SDGs, the targets in this SDG are of a non-AI nature, therefore no further comments are provided here, albeit their fundamental importance for the global sustainable development goals.

The closing SDG 17, related to partnerships for the goals, has five groups of targets: (1) providing finance to the SDGs, (2) technology transfers and funding, (3) training issues, (4) trade issues, and (5) systemic issues related to international cooperation, partnerships and safeguards to the SDGs to be implemented. A special mention is made to the technology targets, providing connections with AmI. For instance, Target 17.6 calls for the international cooperation in all kinds and forms: North–South, South–South and triangular cooperation; regional and international cooperation. Also it calls for the access to science, technology and innovation, and an enhanced knowledge sharing on mutually agreed terms, among nations, with improved coordination by the United Nations, for example. The indicators point to the number of science and/or technology cooperation agreements and programmes between countries, by type of cooperation, and internet coverage for all globe inhabitants. Next, Target 17.7, calls nations to fully operationalize technology banks and science, innovation, and capacity-building mechanism for least developed countries, already past due, by 2017. The main indicator of achievement is also the internet coverage for all world inhabitants. Such targets call for an intense global cooperation with focus on technology transfers among developed, less developed, and poor nations. The issues of AmI are all part of these targets, given their high technology nature.

## 4. Conclusion

The SDGs are unmistakably the most complete and overall framework of collective objectives for humanity to preserve and improve planetary life quality. The proposed actions, stated as "targets", refer to technical and policy oriented tasks. Most of the objectives depend on integrated public policies and international cooperation among nations, institutions and society in general. They represent an intense and present effort towards a global solution for a healthy planet, so far, not achieved by past endeavors.

The SDGs represent the outcome of historic global efforts towards a sustainable world. From a diplomatic standpoint, the first concerns for sustainability were raised in the Brundtland Report, about our Common Future, at the Stockholm Conference, in 1972 [13]. Following, international events took place during the 70's and 80's, but the benchmark for global environment policies was set at the United Nations Conference on Environment and Development (UNCED), called "Earth Summit Conference" held at Rio de Janeiro, Brazil (June 3–14, 1992), that brought high attention to global economic development and environment issues [14]. This meeting created the United Nations Framework Convention on Climate Change (UNFCCC), the Convention on Biological Diversity; a declaration on the principles of forest management conventions for the climate and biodiversity; and the Agenda 21. These conventions and efforts gave birth to intense diplomatic, government and civilian actions towards sustainable development. Nevertheless, most of the advances made, globally and locally, have failed short of the projected objectives, being climate change, the most relevant global issue, concerning the future of the planet.

In the same period that the environment awareness has risen to high concerns, all "human living" issues have equally risen to a high level of attention, such as poverty, hunger, health, education, inequality and racial issues, presently enhanced by the global Covid 19 pandemic. For these reasons, the SDGs were conceived. They were framed into a logical thought of total integration and interdependence of human and nature concerns. No development in human life quality is possible without a healthy and sustainable environment. In this sense the role of Ambient Intelligence technology is key for an efficient environment control, providing the basis for life quality policies to be successful.

The SDGs cover all aspects of human and environmental issues for a sustainable future. In this chapter, a study was made in attention to the various subjects related to natural resources, agriculture, water and sanitation, energy, clean production, urban life, climate change, life below water, forests and all "human" related issues, as key for humanity to continue its path to a better future. Such legacy is implicit in the sustainable development concept, and shall be vigorously pursued by Ambient Intelligence technologies to help the achievement of all targets outlined for sustainable development. In this endeavor, governments, academia, and industries, shall team vigorously to seek extraordinary results in research and development, technology transfers, with attention on the cost and accessibility of these inventions.

The Ambient Intelligence community, should pursue a concerted effort to spread, teach, exchange and foster all kinds of knowledge associated with the SDGs through a careful attention to its targets, clearly expressed in the official UN SDGs guidelines. This study has tentatively selected a meaningful piece of information for AmI professionals and institutions to exchange and explore present technologies and research possibilities in line with the SDGs.

This chapter had the proposal of introducing the key aspects of the SDGs presenting technical and institutional connections with the Ambient Intelligence devices and community. It is feasible to believe that by 2030, when the SDGS are targeted to be in place, Ambient Intelligence will be more present in life and environment checks and controls. Such is the overall legacy of this chapter: a proposition of a growing awareness of Ambient Intelligence devices and institutions, further stimulated to adopt the SDGs as part of their strategies for a sustainable world in 2030. This is an achievable role that can be of relevant importance for the Ambient Intelligence community to prosper and leave a better world for the next generation to come, with strong concern and education for sustainable development, strongly aided by Ambient Intelligence devices and public policies, worldwide.

## Author details

Luiz Cruz Villares
Fundação Amazônia Sustentável (Sustainable Amazon Foundation), Manaus, Brazil

*Address all correspondence to: luiz.villares@fas-amazonas.org

## References

[1] United Nations, Department of Economic and Social Affairs – Sustainable Development. The 17 Goals. https://sdgs.un.org/goals.

[2] Kassai, José Roberto; Carvalho, Nelson; Kassai, José Rubens. Contabilidade Ambiental: Relato Integrado e Sustentabilidade, 2019.

[3] Nordenlund, Lars. Getting Back to Life After the COVID-19 Pandemic with Technology as the Human Factor. https://www.cognize.live/post/life-after-covid-19-pandemic, 2020.

[4] United Nations, Department of Economic and Social Affairs – Sustainable Development. Goal 3, Progress and Info, 2020.

[5] Villares, Luiz, Blockchain and Conservation: Why does it matter. Applications in Payments for Ecosystems Services and Bolsa Floresta Program. Springer International Publishing, Proceedings of Ideas 2019, 2020.

[6] Smith, Joe. This Bird Predicts Hurricanes Better Than Meteorologists. https://blog.nature.org/science, 2020.

[7] Rolnick, David. Tackling Climate Change with Machine Learning – Introduction. Climate Change AI (CCAI), 2019.

[8] Nasa Science. The study of Earth as an integrated system. https://climate.nasa.gov/nasa_ science/science/, 2019.

[9] Lacoste, Alexandre. Tackling Climate Change with Machine Learning – 5. Farms & Forests. Climate Change AI (CCAI), 2019.

[10] The Spatial Web and Web 3.0. What business leaders should know about the next era of computing. Delloite Center of Integrated Research, Deloitte Development LLC, 2020.

[11] Open Climate.Yale OpenLab. https://openlab.yale.edu/open-climate. 2019.

[12] The State of the forests, 2020. Food and Agriculture Organization of the United Nations. fao.org/state-of-the forests, 2020.

[13] Report of the World Commission on Environment and Development: Our Common Future. World Commission on Environment and Development. United Nations, 1972.

[14] United Nations Conference on Environment and Development, Rio de Janeiro, Brazil, 3-14 June 1992. United Nations Conferences, Environment and Sustainable Development. https://www.un.org/en/conferences/environment/rio1992, 1992.

Chapter 7

# Impact of IoT on Renewable Energy

*Sivagami Ponnalagarsamy, V. Geetha, M. Pushpavalli and P. Abirami*

## Abstract

The emerging computing technology in this era is the Internet of Things. The network of intelligence that bridges various devices, systems located in remote locations together by means of cloud portal. IoT maybe equipped with millions or billions of devices. IoT handles large volume of data, process the huge data and performs useful control actions to make our life safe and simple. IoT evolves Human-human communication with thing-thing communication. IoT applications are not confined to a particular sector. In the fields such as health care, smart homes, industries, transportation, etc., the technology which is more influential is IoT. Energy sectors are now undergoing transformation. The transformation is driven by IOT. Green energy without IoT cannot be imagined in this energy sector. Renewable energy sources will be the major power producers among all the other sources due to the depletion of conventional energy sources. Among the renewable energy sources, Solar and Wind contributes more when compared to geothermal, biomass, etc. Renewable energy power production depends on environmental factors such as temperature, wind speed, light intensity etc. These factors affect the performance of energy conversion in renewable energy sources. Since our future generation will depend only on renewable energy, it becomes necessary for the researchers to integrate IOT to provide reliable and affordable energy. Renewable power generation helps in reducing the toxic level of gases which may be produced by thermal power stations during power generation. IoT brings about changes from generation to transmission to distribution. For example, let us compare the traditional grid with that of the smart grid. In the case of traditional one-way communication exists that is power produced from the power station is transmitted to the customer. The customer has to pay for the energy consumed. But smart grid has two-way communication. The customer has the capability to pay for the energy consumed only and if excess power produced can be transmitted to the grid. IoT helps in analyzing the demand as well the wastage of energy, helps in scheduling the load in order to reduce the cost. The sensors and data sciences with IOT helps in achieving the automation and intelligent operation of renewable energy farms, increases the efficiency and reliability of the farms to meet our future power demand.

**Keywords:** IoT-Internet of Things, PV-Photovoltaic, Smart grid, Wind

## 1. Introduction

Renewable energy picks up new pace due to depletion of fuel level at accelerating speed. Energy good for the people as well as the planet is Green resources.

The energy derived from sources which are natural, furnish or renew themselves without depleting the resources of the planet is Renewable Energy. These green re-sources are such as sunlight, wind, rain, wave, tide etc. It can be stated that Renewable Energy is virtually inexhaustible.

Energy sources whether renewable or non- renewable have its own impact on the environment fossil fuel. World still depending on fossil fuel cause climate change by emitting greenhouse gases, also leaves endangered particles that affect the health of the people. Compared to non- renewable the advantages of renewable are it reduces the use of land and water, it also reduces the air and water pollution, no or lower greenhouse gas emissions.

Greenhouse gas of significant are expelled during combustion of fossil fuels but renewable energy emits less or no green-house gases. No or less emission consequence is good climate. Fuel based transport, increase in industrial activities, power generation worldwide increases level of air pollution. Air pollutants make the people to suffocate. World health organization studies confirm that air pollution above urban skies cause premature death. Fossil fuel level apart from getting depleted it also pollutes the environment where as renewable energy meets the objectives that is replaces end of life concept and is seed for social and economic development. Power generation using nonrenewable is influenced more by geo political strife when compared to that of green resources.

Renewable energy cost low and energy prices are affordable. In order to build and maintain facilities more investment is spent for material and workmanship. Renewable energy creates jobs to fuel the economy of the local people because the investment is made within same continents, frequently the same country and usually the same town itself. Renewable energy has lowest cost source for power production and also available to expand. The way available to expand energy access to all areas namely, urban, suburban and per urban. The green resources accessible to all are good signs for development. It is secured and is of good quality.

Renewable energy companies have tremendous global growth last few years. This scaling should maintain profit and productivity. Reports confirm that renewable energy share expected to increase about 22.5% of the global power mix [1]. Use of IoT for renewable energy helps in saving electrical energy. IoT finds out the methods to optimize the capacities of expanding grids across remote locations. Thus, by keeping IoT at the forefront green source companies gain a lot.

Over a last few decade, a drastic change is seen in the energy sector because of upcoming renewable energy and depletion of exhaustible resources. According to International Energy Agency by the year 2022 renewable energy source and IoT economy is expected to reach 43% gain. IoT play a key role in evolving energy sector. It is difficult to envision the future of inexhaustible resource without IoT. Thus, let us look into the possibilities how IoT integrated renewable energy sector meet the growing power requirement as well as how it enhances the efficiency.

## 2. IoT architecture for renewable energy

The layers involved in monitoring and controlling renewable energy sources are perception layer or bottom most layer, network layer, middleware layer, application layer or top most layer. Layer responsible for collecting data is perception layer. It collects detailed information about physical factors for monitoring and managing the environment. The layer is equipped with various sensor devices or other components to meet the configuration to full fill the requirement of user. Next to

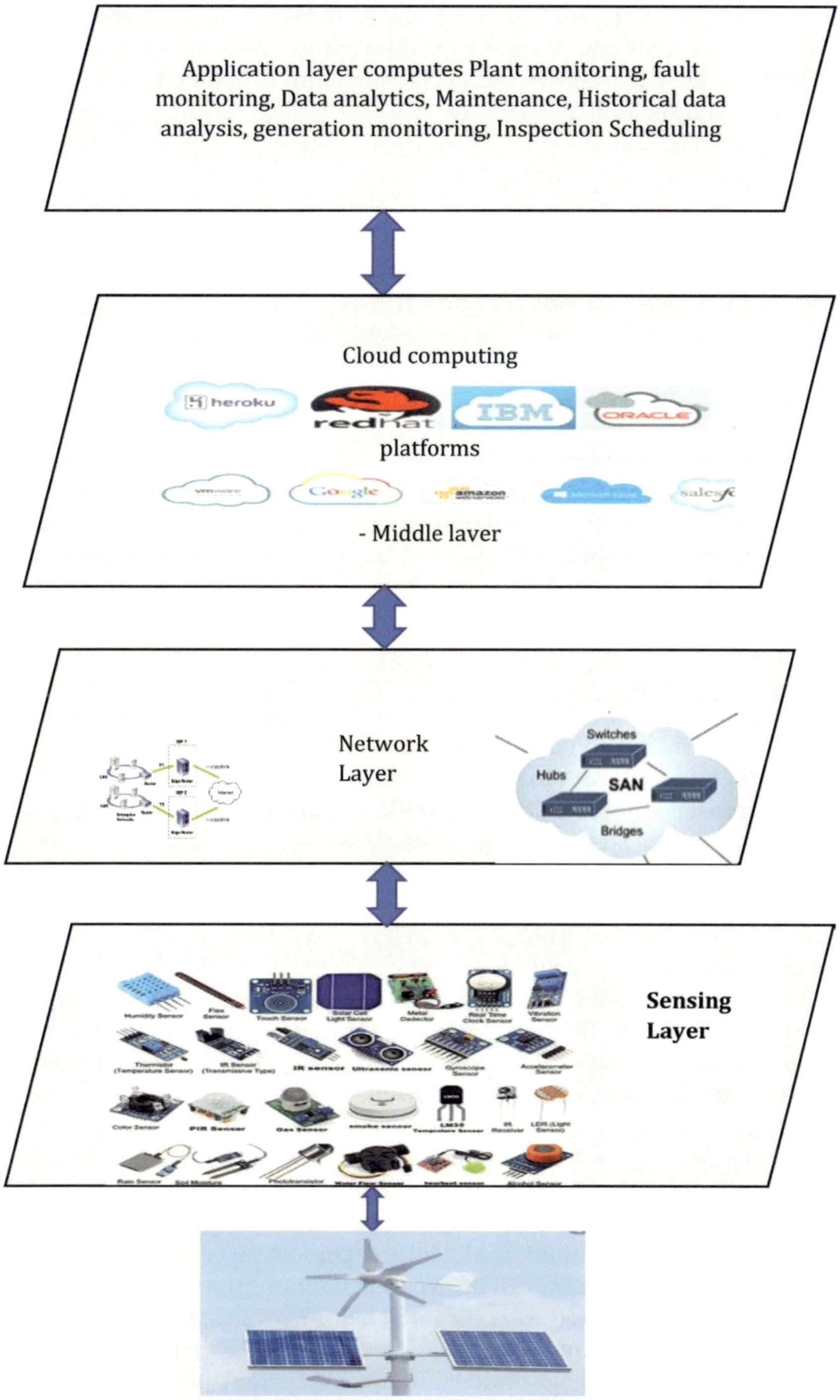


**Figure 1.**
*IoT layers for renewable energy.*

the perception layer is the network layer. It transfers information and data through access and transport network. Access network are short range wireless network like Wifi, ZigBee, sensor area network. Transport network access includes wired or wireless area network. Some of the technologies for transport network access are Internet protocol version 4 and 6, user datagram protocol.

The interfacing of network layer with that of the application layer is carried out by the middleware. This layer extracts the information and convert into required format. Middleware decomposes complex system into simpler one. It follows Service Oriented Approach. Application layer includes cloud computing platform, application platform. The application layer stores the data received from the perception layer via gateway and organizes the data received and computes the data and predict decisions. Thus, application layer delivers gathered information about the monitored parameters to from the web server to the user. The data can be accessed by the user in the form of reports or visual graphs. The layers involved in performing various functions are shown in **Figure 1**.

## 3. IoT for renewable energy

The maximum PV yield is determined by mainly two factors namely unalterable and alterable factors. Unalterable factors are the environmental factors such as solar insolation, wind movement, rainfall, dust properties, ambient temperature and humidity. It has to be accepted by default. Alterable factors allow design flexibility and by varying their design values its response also varies. The alterable factors are orientation of the panel, tilt angle, sun tracking devices, solar reflectors to produce optimum yield.

Physical monitoring of farms such as solar, wind, hydropower etc. becomes difficult and involves human intervention. IoT sensors help in monitoring and managing generation, transmission and distribution across remote locations without involving human interventions. PV set up consists of converter with or without MPPT tracking, controller to feed the grid. Standalone PV system consists of converter with or without peak power tracker, controller, battery storage and inverter Some of the ventures proffered for monitoring the parameters of PV are as follows. The **Figure 1** sketches up the general set up of renewable energy integrated with IoT.

Installation of solar photovoltaic system has seen a rapid growth in recent years because of advancement in technology and reduction in cost of the PV panels. Since PV systems are located at impassable locations it becomes necessary to continuously monitor the performance. IoT based solar PV monitoring uses GPRS module, low cost microcontroller. These components equipped with PV allows the user to access data about the productivity of PV from anywhere. It identifies the fault from the recorded data, informs about the maintenance requirement as well whether yield of PV satisfies the demand or not [2]. A low-cost monitoring system developed using.

Raspberry Pi, microcontroller to provide real time system data in graphical form instead of numerical. Substantial parameters such as voltage, current, environmental temperature are monitored. Based on monitored data decision can be determined whether PV system can be expanded or not [3].

IoT technology supervising solar PV plant helps in evaluating the performance of PV farm in remote location. In this proffered method PV voltage and battery voltage is determined by voltage divider circuit, PV current and battery current is determined by differential amplifier. Grid voltage determined from potential transformer and grid current determined using current transformer. PIC18F46K22 microcontroller as heart of the data logging unit, SIM900 as communication module are interfaced with single core processor AMR926EJ-S. It determines the fault, reminds about the maintenance schedule. Real time data of the plant will make the concerned authority to enhance the decision-making process. Real time data is made available in web server. The remote monitoring becomes essential now a days as the solar farms are interweaved to the utility grid. The energy efficiency of PV farm is improved by IoT based remote monitoring [4].

Agriculture has two major issues namely water scarcity, labor cost. Agribot developed for agriculture purpose is powered by PV. The PV is equipped with boost converter to power the Arduino. Agribot employs YL69 to determine the moisture content of the soil and LM3569 to measure the temperature. The monitored moisture content of the soil and temperature are transmitted to the cloud portal via wireless network. It processes the data and performs control action. The control unit is activated to pump the water to the field when the measured values of moisture content and temperature are below the threshold values. If the measured values are above the threshold limits Agribot just passes that area without irrigating the field. The DC motor is used to drive the Agribot. IoT not only enhances the efficiency of PV but also improves the irrigation of the farm to increase the yield of the farm field [5].

IoT implemented to harness maximum power from PV. PV voltage and current are made available to the cloud portal. The cloud portal processes the voltage and current to determine the actual power. The actual power is compared with that of the set value. If it is equal to zero then the duty cycle for switching device of the converter remains the same. If it is not equal to zero then it compares the value of PV voltage with that of the threshold value. If it is less or greater accordingly the duty cycle of the switching device is varied in order to track maximum power. IoT performance compared with that of different algorithm technique for tracking maximum power and found that IoT converges fast and the implementation complexity is also not high and the parameters considered by IoT are voltage, current and irradiation [6]. IoT replaces the hardware-based stations for measuring the global horizontal irradiance as well as cell temperature and ambient temperature. IoT based module captured the data similar to hardware- based station. It does at a lower cost. Thus, this system provides solar statistics without the intervention of humans and determine in which location sun strike is more and calculates the amount of energy that can be obtained in that area [7].

In Electrical Computer Engineering laboratory at Memorial University 260 watts Solar panel monitored using opensource SCADA system based on IoT. IoT includes the parameters PV voltage and current. The sensors measure the voltage and current of PV. It is made available to the Arduino microcontroller. The values are communicated through Node Red Programming tool. The server platform for storing data, as well as for monitoring and performing control actions is EmonCMS. The experimental setup validated and the outcome of the system proffered holds good not only for monitoring and performing control actions but also for data receiving, transmission and presentation also. The proffered system is suggested for applications such as in generation, transmission and distribution of power, industries etc. [8]. For marine applications Solar photovoltaic system is implemented for generating power and is known as auto generators. In order to determine the energy harnessed by PV it is equipped with infrared camera with 8 megapixel, GPS, transceiver, Raspberry pi, GPS. The images of PV panels are modulated and sent through internet gateway. The modulated images are received using Software Desired Radio and transmitted via ethernet cable for retrieving the actual images at the user end. In order identify the fault from remote location and also to determine whether fault occurred can be recovered or it requires a service. Thus, the energy monitoring device using IoT cutdown the cost and aims at continuous monitoring to harness maximum yield without any interruption [9].

IoT for solar monitoring system proffered uses Amazon Web Services to have a live control over peak power. The performance of the monitoring system is carried out on two IoT platforms Beagle Bone Green and ESP8266. Maximum Power Point Tracker (MPPT) software loaded on these two platforms for buck converter topology. The internet connection performance is determined by two parameters namely perturbation period and its magnitude. The time taken to receive the data of temperature and irradiance, convert analog data to digital data execute the algorithm to determine the output and change the value of duty cycle accordingly must be less than that of the perturbation time period. Both wireless profiles were programmed using python and CPP to determine the runtime of MPPT. The outcome of the proffered system is that MPPT frequency of operation depends on how fast data is transmitted and received. Cloud portal- based methodology for tracking maximum power is feasible [10].

In solar array reliability is one of the crucial factors deciding power generation efficiency. Some of the occurrence which reduce power production is shading which may be permanent or temporary. Temporary is of less severity when compared to permanent which has high severity. The other factors which interrupting the power production are line to line fault, single or double ground fault, open circuit fault, short circuit fault. The way in which IoT identifies the fault is that it executes algorithm technique and predict the reason for occurrence of decrease in yield. Soiling – the aggravation of dust on the surface of the solar panels. It reduces the conversion efficiency. Many cleaning techniques are available. Using IoT the cleaning actions are performed at remote locations without human intervention to increase the conversion efficiency. Moreover, reducing soiling effect increases the life span of the panel otherwise it creates hotspot on the surface of the panel making it faulty. Unmanned vehicle with thermal camera is made to fly over the PV farm. The images are processed using IoT cloud portal to determine the exact location of hotspot on the panel [11–14].

Solar energy harvested fed directly to microgrid to facilitate lighting, battery charging. The computing and control technique provided using IoT. Control technique implemented enhances optimal tracking. This dependable control technique using IoT not only improves the reliability but also the self- recovery of the system [15]. When solar panels are placed on the rooftop it becomes difficult to monitor and study the stability of the system. The sensor units monitoring PV parameters are connected to the Zigbee devices to make the data available to the web. Web server presents the data to the user using Ajax and Html technology. The platform implemented for Zigbee is opensource Linux operating system. Zigbee performance is found feasible for distributed solar cells [16].

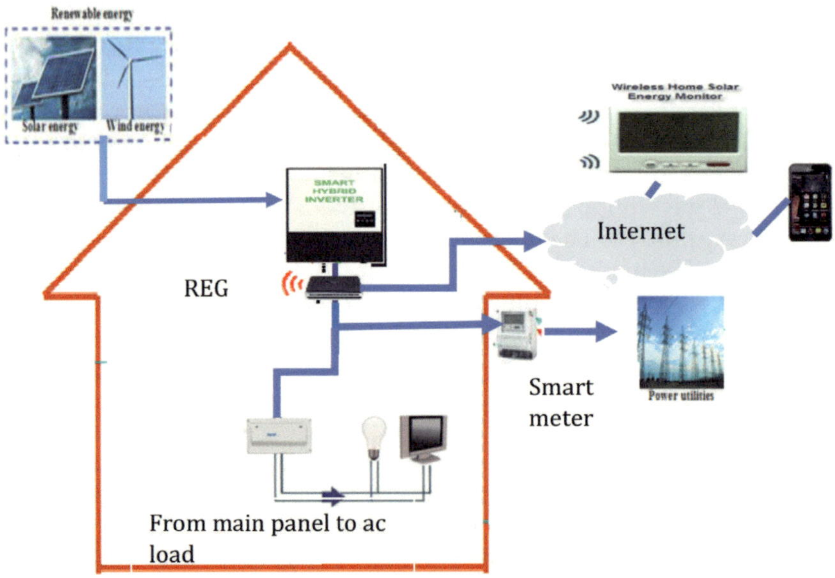


**Figure 2.**
*Remote monitoring for smart home.*

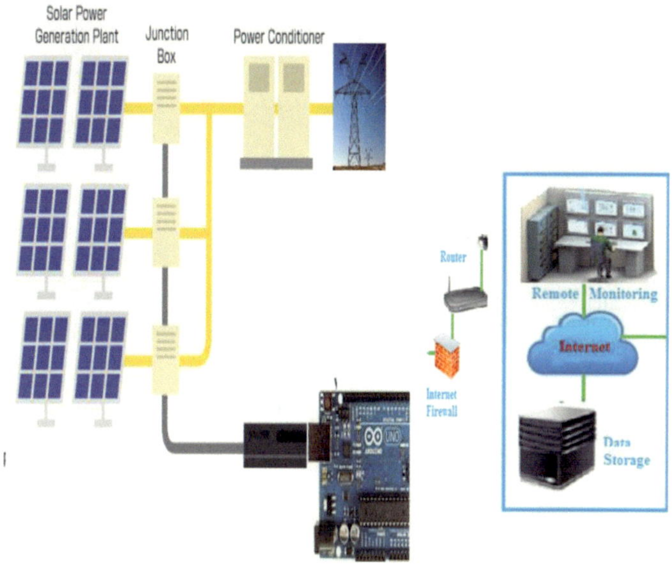


**Figure 3.**
*Remote monitoring for PV grid.*

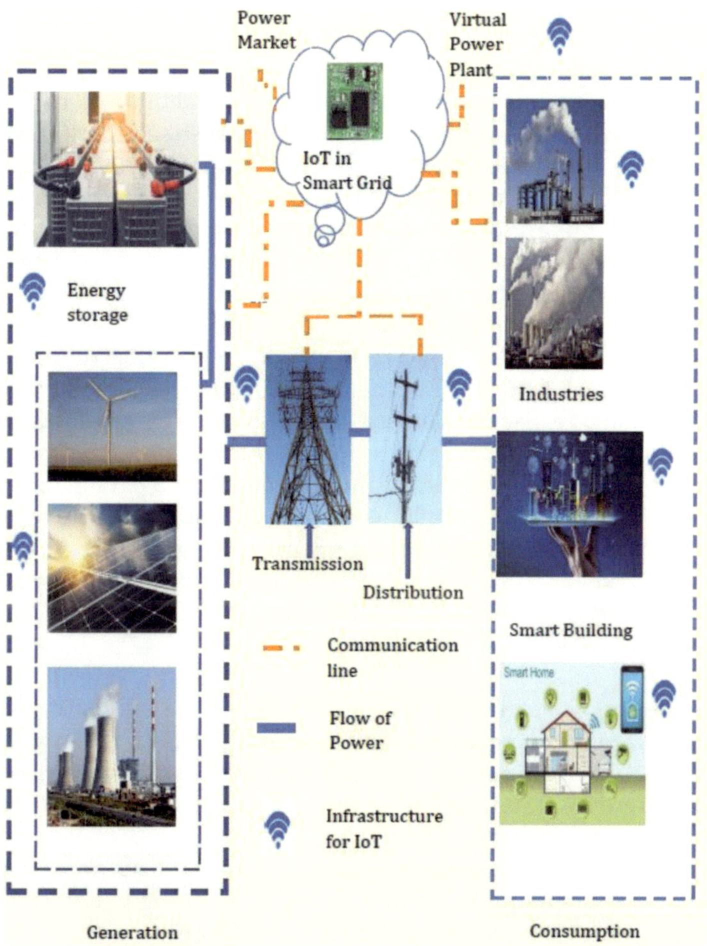


**Figure 4.**
*IoT technology for generation and consumption community.*

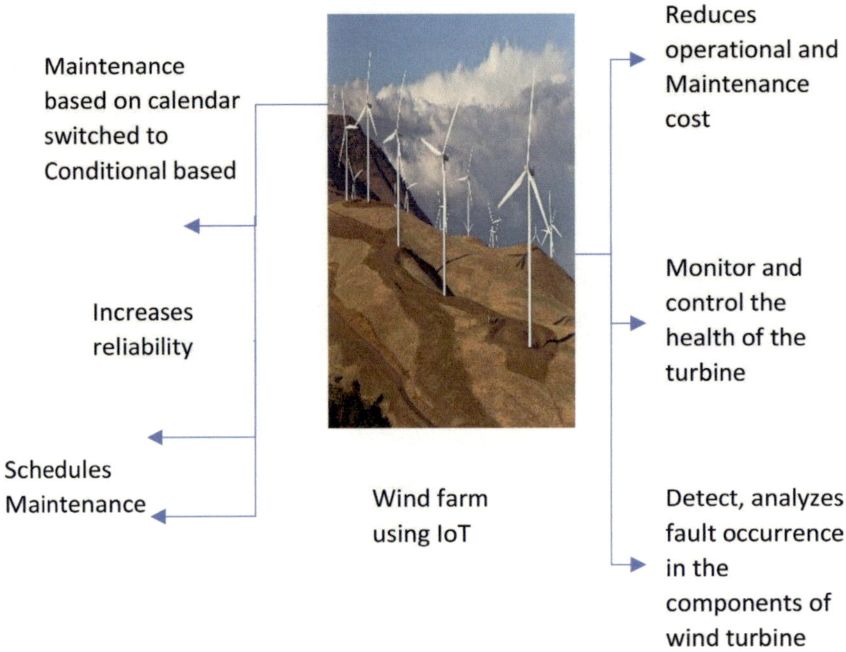


**Figure 5.**
*IoT technology for wind farm.*

Software developed to monitor the parameters PV voltage and current, temperature of the panel, operating time. The proffered system identifies the occurrence of fault automatically and allows remote monitoring using GSM embedded module. With stored data analysis carried out to determine the performance of PV system. Specialist and operators are allowed to monitor the performance. Software developed can be altered or updated according to the requirement by specialist. Operators can just monitor only [17]. The **Figures 2** and **3** delineates the remote monitoring for smart home and PV grid.

The **Figure 5** sketches IoT for wind farm which monitor the parameters of wind turbine such as velocity of wind using anemometer, voltage, current, vibration, humidity, power using respective sensors. The measured real time parameters are made available to the users in remote location via Internet. IoT integrated cloud portal helps to store, analyze the data of measured real time parameters with that of actual data determined using machine learning algorithms. These techniques not only determine the fault but also helps in decision making [18, 19].

## 4. Conclusion

Thus, from the above references in order to enhance the efficiency of renewable energy continuous monitoring becomes necessary which is carried out using various sensor devices. Everything is made simple and easy using evolving technology IoT is shown in **Figure 4**. If renewable energy sources keep IoT at its forefront the benefits it presents are.

Physical monitoring of farms such as solar, wind, hydropower etc. becomes impossible. IoT sensors help in monitoring and man-aging generation, transmission and distribution across remote locations. Continuous monitoring ensures that all equipment work efficiently thus helps in improving reliability.

Renewable energy equipment has complex constructions. They require an adjustment according to environmental change. IoT implemented wind, solar have better control and also helps in reducing operational cost. The advanced monitoring using IoT improves safety on the premises to avoid disaster by providing timely alerts, automatic shutdown etc.

IoT powered smart grid have the ability to detect changes in the demand supply of electricity and is able to react to these changes automatically and provide needed information to manage demand. Thus, improves the energy efficiency.

Energy equipment housed IoT sensors constantly monitor, analyze and predict the energy requirement. The building analyzed da-ta helps in managing infrastructure or to modify the existing infra-structure based on demand. Sensors help in isolating affected area and automatically reroute power as soon as the problem is solved.

Based on accurate measurement using sensors IoT can predict the precise energy generation thus optimizes production and control. The energy consumed reports can be sent to the users so that they can identify the energy wasted and can save energy bills.

Though IoT becomes part of consumers and industries, the problem that revolves around IoT is lack of security. For example, on 23rd December 2015 at Ukraine first power grid cyber-attack took place. The malware disrupted three energy distribution companies. It switched off 30 substations. It left 230000 people without electricity for a period of six hours. Since, IoT devices share information over the internet. Thus, the data hacked affects the privacy of consumers and industries to an extreme level. In order to overcome the issues IoT devices must develop top notch encryption for sharing information over the internet. Apart from this it creates unemployment in developing countries since it replaces human intervention with thing to thing communication. Moreover, people look out to do their task easier with less effort. IoT makes their work simpler and easier at the same time it makes the people lazy and dependent on technology.

In forth coming years IoT with renewable energy transform the energy sector by continuous monitoring, performing control actions to improve the efficiency as well as to distribute the energy harnessed efficiently. IoT implemented with renewable energy saves consumer time as well as money. It reduces the carbon foot print. In future the renewable energy sector will become smarter, efficient and more reliable. Though IoT has many advantages it must put effort to determine the ways in which it can combats its demerits.

## Author details

Sivagami Ponnalagarsamy*, V. Geetha, M. Pushpavalli and P. Abirami
Sathyabama Institute of Science and Technology, Chennai, India

*Address all correspondence to: sivagamitec@gmail.com

## References

[1] Renewables 2020 Global status report; www.ren21.net.

[2] A. Kekre and S. K. Gawre, "Solar photovoltaic remote monitoring system using IOT," 2017 International Conference on Recent Innovations in Signal processing and Embedded Systems (RISE), Bhopal, 2017, pp. 619-623, doi: 10.1109/RISE.2017.8378227.

[3] N. A. Othman, M. R. Zainodin, N. Anuar and N. S. Damanhuri, "Remote monitoring system development via raspberry-pi for small scale standalone PV plant," 2017 7th IEEE International Conference on Control System Computing and Engineering (ICCSCE), Penang, 2017, pp. 360-365, doi: 10.1109/ICCSCE.2017.8284435.

[4] Adhya, S., Saha, D., Das, A., Jana, J., & Saha, H. (2016). An IoT based smart solar photovoltaic remote monitoring and control unit. 2016 2nd International Conference on Control, Instrumentation, Energy & Communication (CIEC). doi:10.1109/ciec.2016.7513793

[5] D. S. Rahul, S. K. Sudarshan, K. Meghana, K. N. Nandan, R. Kirthana and P. Sure, "IoT based solar powered Agribot for irrigation and farm monitoring: Agribot for irrigation and farm monitoring," 2018 2nd International Conference on Inventive Systems and Control (ICISC), Coimbatore, 2018, pp. 826-831, doi: 10.1109/ICISC.2018.8398915.

[6] Shaw, R. N., Walde, P., & Ghosh, A. (2020). IOT based MPPT for Performance Improvement of Solar PV Arrays Operating under Partial Shade Dispersion. 2020 IEEE 9th Power India International Conference (PIICON). doi:10.1109/piicon49524.2020.9112952.

[7] P. Madub, P. Appavoo and A. Chiniah, "Evaluation of IoT for Automisation of solar map production: Case of Mauritius and Outer Islands," 2019 IEEE GlobalConference on Internet of Things (GCIoT), Dubai, United Arab Emirates, 2019, pp. 1-4, doi: 10.1109/GCIoT47977.2019.9058415.

[8] L.O. Aghenta and M. T. Iqbal, "Development of an IoT based open source SCADA system for PV system monitoring," 2019 IEEE Canadian Conference of Electrical and Computer Engineering (CCECE), Edmonton, AB, Canada, 2019, pp. 1-4, doi: 10.1109/CCECE.2019.8861827.

[9] A. Hegarty, G. Westbrook, D. Glynn, D. Murray, E. Omerdic and D. Toal, "A Low-Cost Remote Solar Energy Monitoring System for a Buoyed IoT Ocean Observation Platform," 2019 IEEE 5th world forum on internet of things WF-IoT), Limerick, Ireland, 2019, pp. 386-391, doi: 10.1109/WF- IoT.2019.8767311.

[10] M. Bardwell, J. Wong, S. Zhang and P. Musilek, "Design Considerations for IoT-Based PV Charge Controllers," 2018 IEEE World Congress on Services San Francisco, CA, 2018, pp. 59-60, doi: 10.1109/SERVICES.2018.00043.

[11] Alghamdi Bahaj Blunden and Wu (2019), "Dust Removal from Solar PV Modules by Automated Cleaning Systems", Energies, Vol. 12, No. 15, p. 2923. doi:10.3390/en12152923.

[12] Sivagami, P., Jothiswaroopan, N.M. IOT based statistical performance improvement technique on the power output of photovoltaic system. J Ambient Intell Human Comput (2020). https://doi.org/10.1007/s12652-020-01954-8

[13] Ashley.T, Carrizosa.E and Fernández-Cara.E (2019), "Heliostat field cleaning scheduling for solar power tower plants: A heuristic approach", Applied Energy, 653-660, Doi: 10.1016/j.apenergy.2018.11.004.

[14] Carletti .V, Greco .A, Saggese .A and Vento .M. (2019), "An intelligent flying system for automatic detection of faults in photovoltaic plants", Journal of Ambient Intelligence and Humanized Computing. Doi: 0.1007/s12652-019–01212-6.

[15] Manh Duong Phung, Michel De La Villefromoy, and Quang Ha, "Management of Solar Energy in Microgrids Using IoT-Based Dependable Control," https://arxiv.org/pdf/1710.03422.pdf.

[16] Cheng Yaling, Zou Xiaoping, Zhuang Guangtao, Teng Gongqing, & Qu Weiwei. (2015). Remote monitoring system for distributed photovoltaic based on ZigBee. 2015 4th International Conference on Computer Science Network Technology (ICCSNT). doi:10.1109/iccsnt.2015.7490878

[17] Nalamwar, H. S., Ivanov, M. A., & Baidali, S. A. (2017). Automated intelligent monitoring and the controlling software system for solar panels. Journal of Physics: Conference Series, 803, 012107. doi:10.1088/1742-6596/803/1/012107 s

[18] D. Kalyanraj, S. L. Prakash and S. Sabareswar, "Wind turbine monitoring and control systems using Internet of Things," 2016 21st Century Energy Needs Materials, Systems and Applications (ICTFCEN), 2016, pp. 1-4, doi: 10.1109/ICTFCEN.2016.8052714.

[19] Lina Alhmoud, Hussein Al-Zoubi, "IoT Applications in Wind Energy Systems", Article in Open Engineering, Published November 2019

Chapter 8

# A Panglossian Dilemma

*Christer Johannes Bengs*

## Abstract

Ambient intelligence is a factual phenomenon of increasing magnitude. It also invites intrigued attention as carrier of meanings. Meanings are produced in a variety of contexts, which are here the focus of attention. In order to analyze contextual narratives and their effects, concepts such as intelligence, optimization, rationale, rationality and ambience are discussed. One meaning of ambient intelligence is its indicative contribution to increased unilateral control of the many by the few. Ethical guidelines may be part of prevailing rhetoric, but their success as a self-controlling factor seems fairly unrealistic. Moral confusion is not only related to artificial intelligence, but to the very essence of modern society.

**Keywords:** ambient intelligence, intelligence, rationale, rationality, optimization, ambience, epistemology

## 1. Introduction

Our world is composed of particulars, matters that have extension such as dimension, weight and form. Our lives are also composed of universals, abstractions regarding relative matters such as position and value. Particulars are compulsory to conceptualize when describing the world, universals are indispensable when making particulars and other universals meaningful. Our world exists for us as far as we live, and we live as far as we produce meanings.

Our understanding of the world is constantly deepening due to scientific progress, but the meaning of our lives is not a matter of accumulating knowledge. Every generation and every individual have to work out that for themselves.

Ambient intelligence is a phenomenon that can be described and it is also a carrier of meanings. As rhetoric would advise, the best argument is the inevitable. By skillful descriptions, rhetoric aims are pursued and those subjected to skillful talk eventually think they have figured out everything by themselves. In the pursuit of a critical understanding, conceptual analyses are needed.

Here, an attempt is made to conceptualize contexts that are meaningful for understanding ambient intelligence and the ways we understand it. Intelligence is often associated with the act of optimizing, which in its turn seems to be connected to the broader concept of rational action. Rationality is, however, a function of the context where rational action takes place. The context of rational action has its own rationale, which defines rationality.

Ambient intelligence concerns ambience, and ways we conceive it. In pre-industrial built environments, the physical context caused intelligible ambience to emerge. In the industrialized world of constant flux, intelligible and stable conditions are replaced for a dynamic that makes virtue of the constant need for change.

Whatever that ambience is, it must influence the meanings we attribute to ambient intelligence, and it must have a crucial effect on how societies are managed and controlled by artificial intelligence.

## 2. Intelligence and optimization

What do we understand by intelligence? Obviously, it indicates problem-solving capacity, but what does that mean? The development of intelligence testing is based on their measuring capacity, but what do they measure? An essence of rationality is optimization, but how, and what, can we optimize?

### 2.1 Intelligence

The lexical meaning of intelligence is the ability to acquire and apply knowledge and skills. The ability to acquire knowledge is evidently linked to personal capacity as individuals are born different. Inherent assets may not be realized due to external factors such as malnutrition, deceases, social instability or injuries. The same reasons that hamper people from acquiring knowledge and skills may also cause obstacles in applying them.

The standard definitions seem to pay less or no attention to the potential targets of intelligence. What are the actual contexts where intelligence works? What is the focus? Is it a question of solving particular problems, or to act successfully over longer time periods in changing conditions to achieve some distant end? Does it include only logic and attention, or emotions as well? Is intelligence part of developing and organizing social and symbolic systems?

The more limited the target of our mental activities is, the easier it is to find out ways of optimizing. This is a standard version of rationality. In the gender-centered world some of us still live in, female prejudices attribute "typical male" approaches to "tube-thinking". Male biases attribute "typical female" approaches to "funnel cake-thinking". Either way, rationality is defined according to particular contexts. Males would be accused of lacking the capability to understand matters related to social complexity. Females would face the blame of lacking capacity to rationalize and optimize. This issue exceeds gender speculations as it is an existential matter. All of us are part of the complexity of this world.

The lexical meaning of artificial intelligence refers to the theory and development of computer systems able to perform tasks normally requiring human intelligence. Visual perception, speech recognition, decision-making and translation between languages are often mentioned examples. Optimizing seems to be an integrated and necessary part of programming and the elaboration of algorithms, which makes "tube-thinking" necessary. When expanded into a "funnel-cake thinking", problems arise. Optimization-based rationality gets complicated or outright impossible.

### 2.2 Testing intelligence

Allegedly the first to create a test, in 1905, was Frenchman Alfred Binet (1857–1911) together with his colleague Théodore Simon (1873–1961) [1]. Binet considered intelligence to be a mixture of mental faculties, emerging in changing conditions and controlled by practical judgement. He did not view intelligence as a fixed capacity. Intelligence could not be measured, only classified. The test categorized the mental age of children, and was a way to assess the mental adequacy of the tested compared to the mental average for persons of the same age.

In the USA, eugenicist Henry H. Goddard (1866–1957) got acquainted to the Binet-Simon Scale, and saw it as a way to detect feebleminded people for compulsory sterilization, matching the view of intelligence as genetically inherited. In 1916, Lewis Terman (1877–1956) issued the Stanford-Binet Intelligence Scale, sticking to the view of intelligence as unchangeable.

The pioneer of American behaviorism, Edward Thorndike (1874–1949), defined intelligence in terms of the capability to form neural bonds based on genetic factors as well as experience. J.P. Guilford (1897–1987) maintained that standard IQ tests imply an oversimplified answer, convergent thinking. Creativity on the other hand implies per definition more than one answer to any problem, divergent thinking. He disputed reductionism, and ended up with 180 different types of intelligence, which for practical reasons would limit the use of his method.

In Britten, Charles Spearman (1863–1945) claimed in 1904 that disparate cognitive test scores reflect a single general intelligence factor, and assumed that the psychological g factor would correspond to a biological g factor. This position did not remain uncriticized. Raymond Cattell (1905–1998) developed Spearman's ideas. Fluid intelligence refers to the ability to reason abstractly and perceive relations without previous practice or instructions. Crystalized intelligence generates from experience, learning and accumulated judgement skills. He elaborated a test to assess fluid intelligence by making it culture-fair. His promotion of eugenics has, however, been a cause of critique.

The changing approaches to testing indicate that human intelligence is a controversial matter, and very much embedded in those culture-specific societies from where the theories emerge. Even the fairly recent invention of emotional intelligence (EI) is phrased according to strongly utilitarian guidelines, meaning how to manage emotions to achieve one's goals.

Intelligence testing has historical bonds to biologism and eugenics, which have providing a pseudoscientific basis for racism. Testing reflects the way the overall context of intelligence is conceived. When testing changed from classification to computing, the focus was by necessity narrowed down to matters that could be measured. The perspective should be broadened up as testing intelligence is a moral matter as well. There are many different kinds of utility, and other aspects besides utility to be consider. Is there a happiness-intelligence or only a dissatisfaction-intelligences? Are we looking for creativity, many answers to a problem, or are we looking for optimum, the best answer to one problem?

### 2.3 Optimization

When we optimize, we either seek to minimize resources when pursuing defined ends, or alternatively, we try to optimize results within given resources. Both cases require a time table, often broken down into sub-targets on the way to an end. Optimization may also indicate the attempt to minimize time-use within available resources and defined output, or regardless those.

Economic ventures are typical targets of optimizing, but optimization does not necessarily cover all aspects of a single project. Negative externalities, such a depletion of resources, natural hazards, social and cultural costs that are caused by private entrepreneurs, are still often passed over to public administration and tax payers. In addition, even single projects cannot be optimized without a fixed point of reference in time. In hindsight, many owners of projects would recognize that a change of time perspective could have ended in very different results. An optimum is a function of time.

The issue of benefits to optimize may also be viewed in terms of various kinds of markets according to market access (restricted versus non-restricted) and competition within a market (rivalrous versus non-rivalrous). The market for private goods is per definition restricted and rivalrous. One can enter only in case demanded resources are possessed. Optimization is possible and needed for private benefits. The idea of an unrestricted and free market is an abstraction as the very logic of capitalism induces market restrictions and monopolies. If not, there would be no use for anti-trust legislation. Governments and politics can influence the market of private goods mainly indirectly, by implementing laws and regulations.

Club goods indicate restricted and non-rivalrous markets, which the club can optimize according to conceived club-benefits. Markets for common goods are non-restricted, but rivalrous and the common assets are at risk of being depleted, i.e. the "tragedy of the commons" [2]. Because of non-restrictedness, an optimization is impossible, and public government can interfere only indirectly. Public goods are open for all and do not imply rivalry among users. Because of their open access, there is nothing to optimize from the point of view of public government, except for goods that have to be produced and managed. Sunshine is free for all, but public space needs to be built and maintained.

Singular optimizations sustain competition and the destruction of competitors. But what about the overall economic system and the wellbeing of citizens? Optimizing parts may cause an overall disastrous waste of resources. Adam Smith (1723–1790) claimed there is an overall order in the chaos [3]. He proclaimed that the totality of self-interested actions would eventually cause unintended social benefits. A prudent reader may recollect that the "invisible hand" of markets was not all that invisible: Smith worked for the monopoly at the time, the East India Company.

Governing the national economy is now executed according to the same logic as single ventures. It is boiled down to a restricted number of indicators, like the GNP, and aims at optimizing economic growth. Growth is an end in itself, and the focus of public and general interest. In political rhetoric, positive as well as negative growth lend themselves to very far-reaching conclusions as to their alleged effects on human matters.

GNP reflects the sum of its constitutive parts, which are thought to be optimizable. Nonetheless, a considerable part of the economy is no target for optimizing at all. Common and public goods, being related to public interests such as the smooth running of everyday life, care for tax-payers money and public revenues, are optimized by the political system. The "political system" is a very vague term that may reflect anything from particular interests to the whole body of citizens, or even to humanity as listed in human rights. Insofar as politicians optimize their commitments, they usually focus on the lengths of their tenures.

Human intelligence seemed to escape us, but so does artificial intelligence! For the majority of people, GNP and its annual fluctuations is a very poor indicator for quality of life. Nor does the investor-driven use of artificial intelligence for programming maximum revenues at the stock exchange say much about the utility of the exchange for citizens in general. Maybe the question to ask ourselves is not how artificial intelligence can be humanized, but rather why human life has been reduced to forms that can be optimized by artificial intelligence?

## 3. Rationale and rationality

To conduct oneself intelligently in a rational manner, one has to relate one's actions to a given context. What is the context and how is it formed? Is it something to be made up from case to case, or is it more general? Does rationality change according to context? How to choose when one has to? Does choice by necessity indicate moral judgement? What is the role of science in all this?

### 3.1 Rationale

Rationale refers to controlling principles of opinion, belief, practice, or phenomena. To be rational refers to having reason or understanding, or to something being agreeable to reason. Controlling principles are not perforce agreeable to reason as they may be structural and unintended outcomes of very complicated social processes. Nobody can escape being bound to some sort of overall principles of action, but few can claim to act rationally in every instance.

Dr. Pangloss is a stunning character in Voltaire's novel Candide, published in 1759 [4]. Voltaire (1694–1778) is thought to have used the character for ridiculing Leibnizian optimism. Nonetheless, Dr. Pangloss certainly makes sense as a representative of the breaking times when the traditional teleological world view - the purposefulness of everything - had to confront a causal world view, based on science. But Dr. Pangloss is more than a caricature of naïve optimism, he mirrors an existential dilemma as well.

According to the doctor, "all is for the best", because we live in "the best of all possible worlds". God is the ultimate good so why would not his creation be the best as well? Thus, it is reasonable to claim that everything that occurs is for the best. Dr. Pangloss firmly professed causality within an overall scheme of teleology, thereby reflecting a view of God as the Creator, not as the Intervener. At the time, the existence of God was not questioned, but his nature was.

A problem with Pangloss' ethical position is that everything turns out both acceptable and obligatory, in accordance with the initial ruling of the Creator. It is not Pangloss' fatalism that gives rise to moral doubts, but his opportunism. Actually, his character may be seen as an embodiment of alleged Jesuitical sentiments: End justifies means! If the initial creation is the best of all worlds, then every derivative of that creation, good and bad, is for the eventual good. Only human shortsightedness would blur that post-factum.

As final explanations, the concepts of cause and purpose may appear to us mutually exclusive. But, if we define the purpose of our universe to be causal, there is no contradiction. If the purpose of the universe is defined not to be causal, a contradiction arises. Consequently, to be considered rational we have to avoid thinking and acting in a way that would offend the rationale of our basic guiding principles, whether religious, atheistic or agnostic. Human characters who possess the quality of not being self-contradictory, are thought to have integrity.

We may face another problem as well: What are those entities that generate controlling principles of opinion, belief and practice? Dr. Pangloss was a character of a firmly Christian country of Christian Europe. In a hierarchical manner, any entity can of course be thought of as being part of a greater totality. The Christian solution is to close the hierarchy by referring to this world, the Creation, as the target of human reasoning. The Heaven or Paradise are per definition out of reach, and conceivable only as part of eternity, and so are our understanding of the deeds of the Lord. Any endeavor to bridge the gap may provide ample room for speculation, accompanied by a never-ending stream of self-promoting prophets and wizards.

The Christian world view is by no means unique, rather the contrary. Most of us seek - consciously or unconsciously - to build our identities based on some kind of view of a world that we can and want to live with. Are we free to choose? The gospel of the modern world is: Yes! In reality, experience transmits a more complicated story. Only madmen are able to extrapolate their madness into the big world. The sane ones must go the other way around. Societies and cultures provide rationales, the task of individuals and single ventures is to provide matching thoughts and deeds.

### 3.2 Rationality

In his Utopia, Thomas More (1478–1535) sought to find a rational, explicit and measurable expression for the rationale of Christian society [5]. He was decapitated by his King, Henry VIII, who usurped the religious power of the Pope, and robbed the Catholic Church of its wealth. Maybe the modern world was born in 1535 CE? What are the fundaments of our modern world? Heaven got lost because eternity got lost. Now, our haven (short of the e) is located in this world, but in the future. Remarkably, the end was changed, but the idea of Christian eschatology is still there.

The first to make the switch were the people of the Renaissance. They started to look ahead by looking back. Nonetheless, they applied a conception of time that was linear, albeit opposite to ours. The great discoveries of the early modern time brought about global trade, and in its wake, colonial subjugation, looting and plunder of the Americas, Africa and the East. Economic wealth in Europe brought about a surplus that was reinvested for the sake of further surplus. The future in this world was eventually found.

The corporate form of capitalism that emerged during the 17th century, indicates a rationality narrowed down to optimizing the revenues of single ventures [6]. Over time, some part of the aggregated surpluses has been invested in political ventures labelled charity, corruption or money laundry according to prevailing conjuncture. Concentration of wealth caused by necessity the need for controlling politics, which is now equally obvious in democratic and nondemocratic countries.

During the Renaissance, Antiquity was thought to represent the ultimate achievements of mankind. Social progress is an idea of the 17th century, but the concern was limited to the economy [7]. Towards the turn of the century, a debate in the French Academy between the "Moderns" and the "Antiques" reflected a broader understanding. The issue at stake was the very essence of change: Is all change for the better? After decades, a reasonable conclusion was reached: Quantifiable knowledge can be accumulated, like mathematics and science. Knowledge involving judgement like questions regarding moral and beauty, are skills that individuals acquire, and the knowledge of those cannot be accumulated [8]. There is an endless growth of applicable criteria for making judgement, but that does not indicate improved quality of factual judgements.

Only the Enlightenment of the 18th century, with Voltaire and others, brought to the fore a notion of overall progress, and Dr. Pangloss became a ridiculed figure [9]. He was stuck to the eternal heaven, not the haven of the future. During the heydays of the Enlightenment, progress turned limitless as well as endless, and a purpose in itself. Consequently, the 19th century brought with it progress and regress as ideological and political concepts. In the 20th century, when progress was boiled down to economic growth as indicated by GNP, every economy of the globe could be integrated into a common ranking list with regard to overall output per year and person.

The eventual point of reference is the future of this world. Nevertheless, like the gospel, the future is unverifiable. But it is an offer one cannot refuse as there is nothing to lose, only to gain - except for infidels refusing to give up their integrity. There is a difference between eternity and the future in that the future is even more abstract than eternity. As the case of More shows, his utopia was firmly anchored in Christian ethic. Considering history, it is hard to discern how our future, being a battleground for ideologies and countries of all shadings, has anything to do with particular moral sentiments or ethical considerations.

However, even the haven of future may have an end. When most aspects of human life are increasingly bound up to external order and control, the prospects of single individuals are narrowed down. Now, the wealthiest 10 percent of the global population owns 81.7 percent of global wealth, and the wealthiest 1.0 percent have 45 percent [10, 11]. What happens when 0.1 percent of the global population will own everything? The future could then be not to gaze into the future, but to return to the initial state of human history of here and now. Carpe Diem, catch the day!

The nucleus of wealth accumulation is now finance. The value of money, when being a commodity exchanged on a market, is subjected to fluctuations determined by supply over demand. With concurrent fiat money, the logic changes insofar as investments do not by necessity concern productive measures at all. Finance becomes a club good. By the financial transactions of the biggest players, the value of existing wealth can be manipulated for the sake of more wealth. When the total amount of indebtedness grows faster than productive output, a further concentration of wealth to the club members seems inevitable. A recent estimate suggests a global debt burden of 272 trillion USD, that is 365 percent of total GDP [12].

Rationality seeks its rationale among available possibilities. In the various phases of human development, options at hand may have increased in absolute terms, but they may decrease further in relative terms. The employed criteria of judgement may still expand and improve over time when based on expanding sets of data. The quality of judgement is up to prudence. Individuals are prudent, not nations, and judgement skills can be improved only during a lifetime.

### 3.3 Moral choice

For half a millennium, European science has been developed to encompass most aspects of life, but still there seems to be no theoretical consensus on judgement. In order to make a judgement, one needs criteria, but to figure out criteria, one needs to make judgements. The idea of "value" is self-referential. To evaluate, we need to evaluate and choose applicable criteria, in absurdum [13]. All of us have to make choices, no matter how informed we are. Most choices are moral ones and based on considerations about right or wrong. Moral considerations are not always manifest, but unavoidable and omnipresent.

The Sisyphus-work of redesigning morality is manifest in the ways scientists and philosophers have tried to grip the task. The initial phase was filled with optimism. The grand utilitarian, Jeremy Bentham (1747–1832) aspired in vain to elaborate a felicific calculus, but it would not have included "natural and imprescriptible rights", which he considered "nonsense upon stilts" [14]. His position is rational as utilitarianism was embedded in the economy and politics of his time. The recognition of human rights would certainly have been obtrusive as human labor was supposed to be a commodity of the marketplace.

John Stuart Mill (1806–1873) expressed the idea that the rules of thumb of everyday morality would get endorsed by the systematic utilitarian method, but such derivations are still on their way. The futility of expecting a feasible algorithm of moral values for global cost–benefit analysis is as obvious as ever before. Utilitarian calculations face many problems. Considering positive effects as benefits seems to be obvious, but what about negative effects? In the short run they are costs, but in a longer perspective they may turn out to be beneficial. By switching the perspective, short term positive effects may later on turn out to be negative.

In all, to judge and weight all moral consequences in terms of benefits and drawbacks is impossible. Moreover, even to weight practical results in terms of benefits and drawbacks is impossible, except for limiting the scope to a short period of time and a narrow place. This means utilitarianism reflects a rationality that is conceivable only within the clearly defined limits of single projects.

The Kantian tradition - stressing principles of conduct - has likewise paid tribute to practicality, and resented the impracticality of utilitarianists. The maxims, such as the Categorical Imperative, are open in a similar way as the utilitarian endeavor for benefits. They require an actor to consider and select relevant maxims to match actions or to select relevant actions to match maxims. A truly thoughtful person may not be able to take any actions at all as uncertainty is our companion.

A somewhat sloppy conclusion would be that sincere moral thinking requires understanding, knowledge and imagination, which is not achieved by applying formulae. The complexity of real-world problems is impossible to compute. We can never consider all things, or all times for that matter. In practice, capitalism, and to some extent representative democracy, mostly set a time front that is as long as an investment period or political tenure. Those may be optimized. The positive and direct effects, and alleged positive externalities, are annealed while negative externalities are easily unrecognized or silenced.

Is there a single point of departure, one perspective from where to assess ideal rationality? The traditional answer is yes, common interest. In practice, hardly any political party would miss to refer to public or common interest. The idea of a common interest is illustrated by the Prisoner's Dilemma [15]. To optimize his situation, the rationally acting suspect would judge his fellow suspects and probably find out that some of them are somewhat irrational, and therefore unreliable. The shortsighted self-interest of some accused would obstruct the possibility to find an optimal solution, common for all. Consequently, the ideally rational player would have to turn less rational, not to lose too much. Is that rational? Nonetheless, it seems to be part and parcel of politics, rhetoric and modern life.

### 3.4 Accumulation of knowledge

Scientific institutions worldwide try to safeguard the academic virtues in order to contribute to the accumulation of knowledge [16, 17]. This can be seen as a moral prerogative for science and its global body of researchers. It is also an example of the match between the rationale of science and the rationality of academia. The academic routines include dissertation and publishing of findings, peer reviewing and critical scrutiny, acceptance to prove or disprove arguments regardless the status of the speaker, demand for theoretically anchored hypotheses, reliability of data, application of credible methods, inherent logical consistency of the work, willingness to rework one's findings, etc.

With the increased strategical and commercial impact of science, such traditions are evaporating for the sake of circumscribing and monopolizing the use of knowledge. This is particularly true for breaking research in technology and big pharma in closed institutions, where foreseen benefits are astronomical in terms of revenues and strategic power. In absolute terms, scientists may be more and more knowledgeable, but in relative terms, the opposite prevails as research and development is out of reach for the public, and for most researchers as well.

## 4. Ambience

The lexical definition of ambience is a feeling or mood associated with a particular place. Environment is a token of history, and an analysis may bring understanding of the rationale that drives the present development of ambient intelligence. Firstly, ambience relates to perceived integrity of the environment, but in what sense? Secondly, what changes are obvious when comparing the way production of modern urban environment is organized compared with the traditional ways of building and planning? Thirdly, how does urban form indicate the rationale of economics as well as social and political control?

### 4.1 Traditional integrity

Differences in ambience usually play out to the advantage of historical settings. This is not only a matter of opinion, but reflected in the concept of gentrification, which indicates the preservation and upgrading of historical urban settings, and associated with an influx of new inhabitants and soaring real estate prices [18]. Much of travelling and tourism is based on the fact that historical environments offer a kind of ambience that modern urban settings are void of [19].

Why are historical urban environments so sought after? Why do they please people? One reason is that they associate to important historical events, which are integrated into nationalistic rhetoric. A feeling of nostalgia is probably globally present in the sense that it may remind us of childhood, passed times and our identity.

However, there is another and more tangible reason for the attractiveness of historical urban settings. They are results of handicraft, built out of local materials, following local building traditions, erected by local labor force, which generate overall unity. The finest of historical buildings have pursued a very long life [20, 21]. Representing handicraft, traditional architecture possesses an additional quality. Details of buildings are to some extent distinguishable at a distance shorter than 300 meters [22]. When one approaches them, new and smaller details unfold at closer distance. Handmade environments offer continuously new excitement for a pedestrian despite the fact that she or he may have lived in the surrounding for decades.

The first cities known to history were built in a way reflecting the rationale of tribal society. Each group and segment of the local society managed and controlled its own territory. The first European cities breaking this pattern were the Greek cities of the Antique at the time when the city states and citizenship emerged. Those cities were unlocked in the sense that all parts except the privately controlled plots became available for the citizenry. Houses continued to be produced by the inhabitants for their own purposes. Plans were laid out in advance and lots were distributed by means of negotiations and consent, not as commodities exchanged on a market. Ideally, the control was executed in a communal way by the citizenry for the citizenry [23].

The earliest indication of the idea of landed value is a map of central Florence of the early 15th century, showing the taxation value of properties [24]. At about the same time, the central perspective was introduced into visual arts as a new innovation. Both of these phenomena indicate a novel way of distancing oneself. The use value of the physical setting acquired an additional exchange value. The central perspective provided the viewer with a position that used to be reserved for celestial figures and the Omnipotent. Economic and visual alienation seem to have occurred in correlation.

The relation between the citizenry and the ruler remained in some sense reciprocal. Even in the case of the Baroque city plans of the 17th and early 18th century, the people had visual access to the palace of the Prince who likewise could see every corner of the city from his palace. The religious justifications of worldly inequalities did not diminish the need for overall community. The ambience of historical urbanism expresses integrity.

### 4.2 Modern disintegration

The birth of modernist architecture coincided with industrialization of construction. The pioneers designed their works in a style mimicking the design of factory produced items, although the buildings were produced by handicraft [25]. An argument that has been reiterated over and over again concerns integrity of architectural expression. Modernists claim that architecture has to be honest [26]. As honesty is a relative matter, it has to be related to something. The true point of reference for modernists is time, the spirit of our time, heading for the future – whatever that may indicate. The true expression of any era can be confirmed only in hindsight, which would disqualify the assumed spirit of the present and the future as intelligible points of reference. We cannot pretend, if we want to be truthful!

By associating architectural expression with the modern rationale of continuous reinvesting and rebuilding, the destruction of historical settings became acceptable and even preferable. The place, locality and history lost their meaning as points of reference for determining environmental values. Integrity is understood in terms of the future, not in relation to the past and the actual place with its local characteristics and traditions. Consequently, modernistic urban settings and architecture have no homeland, and built environment is globally uniformed – like artificial intelligence.

There has been some opposition to these trends, for instance a quest for genius loci, the spirit of the place, for topophilia and for critical regionalism as opposed to global design [27–29]. The results are close to neglectable, and do not exceed a limited number of hailed examples. Postmodernism as architectural style is sometimes associated with anti-modernism, but more so it is another expression of modernism. Various approaches that could be summoned under the concept of retro, are also modernistic in the sense that they are integrated parts of modern settings in constant flux, whether exterior or interior.

The Baroque era still expressed reciprocity between controllers and the controlled. This changed only in the late 18th century, when Jeremy Bentham, the utilitarian, introduced the so-called Panopticon for correctional institutions [30]. Due to the design of the precinct, prisoners were constantly surveilled by the guards, who themselves were invisible to the prisoners. Societal control became unilateral. No wonder Bentham ridiculed natural and imprescriptible rights. In a context of unilateral and total control, there can be no room for any inherent right of the subdued, and benefits are much easier to calculate when they concern only those in command. All concurrent systems for urban surveillance are based on the Panopticon principle. Humans are replaced for a huge variety of surveillance technologies, exempt from the controlled.

Planning legislation of the 19th century was still based on the presumption that plot owners would exploit their property for their own needs. In case of purely speculative projects, a developer would have to stick to approved town plans and available plot supply [31]. A century later, planning legislation was turned the other way around to suit large scale speculation in rising land values. Despite the existence of public planning monopolies, developers acquired the right to develop land much as they pleased [32, 33]. The development of planning legislation in Sweden is a case in point. Planning is in practice removed from the public to the corporate sphere and made a club good.

Consider the overall shape of urban environment. Historical cities produced in a traditional way, express an endless variety within an overall unity. This is likely to be the most important single factor that makes historical environments so attractive. That is their ambience. Modern settings express the opposite: Monotonous labyrinths within an overall chaos. Consequently, orientation and identification are made almost impossible, and the best, if not only way to orientate is to use electronical equipment for navigation. That is certainly a need of today, but it is a previously unknown need that did not exist when human habitats were laid out in an intelligible way.

## 5. Ambient intelligence

Ambient intelligence is described by providing general outlines and jots of self-criticism, which set the agenda for further discussions [34]. That is not exceptional, but is it credible?

### 5.1 The phenomenon

Ambient intelligence refers to environments that are sensitive and responsive to the presence of people by means of electronics. In harmony with the modern view of our future haven, it was developed as a corporate initiative in the late 1990s to provide a projection on the future. Information and intelligence were supposed to be hidden in the network that connected different devices. The technological framework behind them was thought to gradually disappear into the surroundings until only the user interfaces remain perceivable by users. The parallel to the Panopticon way of unilateral control is striking!

The ambient intelligence paradigm builds upon computing, profiling, context awareness, and interaction design. Applied systems and technologies are supposed to be context aware as they recognize individuals and their situational context. Moreover, they are personalized and tailored for individual needs, and adaptive as they can respond to individuals. They also anticipate individual desires without conscious mediation. The parallel to an age-old narrative, the life of the master and his servants, is obvious.

Ambient intelligence is said to rely on user experience, and the advancement in sensor technology and sensor networks. In response to operational obstacles, a design emerged that created new technologies and media around the user's personal experience. The user is asked to give feedback to improve the design. Biohacking may be an example that illustrates the most private sides of such applications, which seem to draw the line between private and public inside the body of the users.

Ambient intelligence requires a number of key technologies to exist. These include unobtrusive, user-friendly hardware and human-centric computer interfaces. Computing infrastructure is characterized by interoperability, networks and service-oriented architecture. Systems and devices must be reliable and secure, achieved through self-testing and self-repairing software and privacy ensuring technology. The promises for the future resemble those of salvation of the afterworld.

### 5.2 Criticism

It is said that any immersive, personalized, context-aware and anticipatory characteristics bring up concerns about the loss of privacy. At the same time, it is claimed that applications of ambient intelligence do not necessarily have to reduce privacy in order to work! In social sciences, the possibility of flaws is a question of probability. Nuclear accidents and related catastrophes offer a realistic analogy. According to safety calculations, nuclear disasters would never happen, because the computed probabilities are neglectable. They still happen! Intrusion is an everyday phenomenon, and it is difficult to imagine that hacking would decrease when information systems expand and get more complicated and difficult to guard.

Power concentration in large organizations, a fragmented, decreasingly private society and hyperreal environments where the virtual is indistinguishable from the real, are said to be the main topics of critics. But what about the sector as a main factor in the general tendency of concentrating wealth and power? What about the major global technology companies, accountable only to themselves? Should not that be addressed as well?

### 5.3 The Santa Claus' list

According to the Information Society and Technology Advisory Group (ISTAG), the following characteristics will permit the societal acceptance of ambient intelligence: Ambient intelligence should facilitate human contact, be oriented towards community and cultural enhancement, help to build knowledge and skills for work, better quality of work, citizenship and consumer choice, inspire trust and confidence, be consistent with long term sustainability—personal, societal and environmental—and with lifelong learning, be made easy to live with and controllable by ordinary people [35].

Consider the global social media platforms of today, applying the principle of unilateral control. Now, literally billions of people produce information about themselves, free of charge, to be sold by gigantic operators to other corporations and public authorities. It is surveillance of a magnitude that used to be unimaginable. Here, the essence of artificial intelligence is exposed. It may provide benefits and joy for the billions while enriching global corporations, tightening the straitjackets of ordinary citizens and providing the database for individualized control as well as manipulation of consumer choices and political commodities [36]. The Santa Claus' list appears equally important and naïve.

## 6. Conclusions

It is easy to laugh at Dr. Pangloss' assertion that our noses are shaped to carry spectacles, therefor we use spectacles. But concurrent designers of spectacles may actually think like the doctor, and so may programmers as well. Designers and programmers are professionals, and the rationale of professions is that they reserve for themselves the right to judge what is accountable knowledge. In their practice, evidence-based knowledge and professional judgement are not necessarily kept apart. Drawing up a list of all the good things ambient intelligence should promote resembles Dr. Pangloss' explanation why his friend drowned in the bay of Lisbon: The bay was created for that purpose!

An obvious parallel is the tenet of business that economic growth must be pursued for the sake of economic growth, because in the best of worlds there is perpetual economic growth. Technological development is of course a constitutive part of that narrative. That part also includes the (professional) presumption that ethical guidelines are a matter for the sector itself. MIT professor, Dr. Tegmark has pointed out the urgent need for ethical guidelines, elaborated by the sector itself [37]. Kindly expressed, he cannot be familiar with avalanches of financial disasters, instigated by the financial sector for some centuries now, under the auspices of self-regulation.

The fundamental dilemma is not whether to promote ambient intelligence or not. It will be developed anyway. But how to work out ethical rules that would safeguard users from intrusion, fraud, blackmailing, trafficking, abduction of identity, robbery, or commercial, social and political manipulation, or global surveillance of each and every individual – all the horrors of Pandora's box?

As far as ethical rules are concerned, the problem is not only related to artificial intelligence, but to the very essence of modern society. We are living in a world in constant flux, where uncertainty is said to be increasingly replaced by rational decision making, backed by science and new technology. In the best of worlds, that process would eventually make individual judgement and moral choices obsolete. However, we are not quite there yet, and the outspoken idea of modern societies is not to be judgmental. The contradiction between ideology and reality indicates a vast grey zone, where Pandora's box is wide open. Voltaire and Dr. Pangloss may have died, but the Panglossian dilemma lives!

## Author details

Christer Johannes Bengs
Swedish University of Agricultural Sciences, Sweden

*Address all correspondence to: christer.bengs@aalto.fi

## References

[1] Benson, N. et al. The Psychology Book. New York: DK Publishing; 2012. 352 p. ISBN 978-0-7566-8970-4

[2] Ostrom, E. Governing the Commons. The Evolution of Institutions for Collective Action. Cambridge: Cambridge University Press;1999. 280 p. ISBN 0521405998

[3] Smith, A. An Inquiry into the Nature and Causes of the Wealth of Nations. London: Penguin: 1999. 602 p. ISBN 0140436154

[4] Voltaire. Candide. London: Penguin; 1997. 163 p. ISBN 0-14-062263-2

[5] More, T. Utopia. London: Penguin Books; 2012. 146 p. ISBN 0141442328

[6] Ferguson, N. The Ascent of Money: A Financial History of the World. New York: Penguin Press; 2008, 441 p. ISBN 9781594201929

[7] Heller, A. Renaissance Man. New York: Schocken Books; 1981. 481 p. ISBN 0805206744

[8] Lowenthal, D. 1995. The Past is a Foreign Country. Cambridge: Cambridge University Press; 1985. 489 p. ISBN 0521224152

[9] Kant, I. An answer to the question: "what is enlightenment?". London; Penguin; 2009. 120 p. ISBN 9780141043883

[10] Credit Suisse. Global Wealth Databook: 2019. Available from: https://www.credit-suisse.com/about-us/en/reports-research/global-wealth-report.html

[11] Bengs,C. Marcet Sine Adversio Virtues. In: Bevilacqua et al. (Eds): NMP 2020, SIST 178; 2021. p. 396-405

[12] IIF 2020. International Institute of Finance, The Institute of International Finance > Research > Capital Flows and Debt > Global Debt Monitor (iif.com). Available from: https://www.iif.com

[13] Bengs, C. Pluralitas non est ponenda sine necessitate. In: Procedia – Social and Behavioral Sciences 223; 2016. p. 561-567

[14] Dennett, D.C. Darwin's Dangerous Idea. Evolution and the Meaning of Life. London: Penguin Books; 1996. 586 p. ISBN 014016734X

[15] Peterson, M. (ed.) The Prisoner's Dilemma. Cambridge: Cambridge University Press; 2015. 298 p. ISBN 9781107621473

[16] Bacon, F. The New Atlantis. [S.I.]: ReadaClassic com. 2010. 51 p. ISBN 9781451537635

[17] Wootton, D. The Invention of Science. A New History of the Scientific Revolution. New York: Harper; 2015. 769 p.

[18] Lees, L. et al. The Gentrification Reader. New York: Routledge; 2008. 310 p. ISBN 9780415950367

[19] Park, H-y. Heritage Tourism. London: Routledge, Taylor & Francis Group; 2014. 251 p. ISBN 97804115595827

[20] Benevolo, L. 1980. The History of the City. Cambridge: MIT Press; 1980. 1011 p. ISBN 0-262-02146-3

[21] Morris, A. E. J. History of Urban Form Before the Industrial Revolution. Harlow: Longman Scientific & Technical; 1994. 444 p. ISBN 0582301548.

[22] Benevolo, L. Fixierte Unendlichkeit: die Erfindung der Perspective in der Architektur. Frankfurt/Main: Campus Verlag; 1993. 144 p. ISBN 3-593-34830-6ISBN

[23] Bengs, C. 2018. Perspectives in Perspective – Seeing, Visualising, Ruling (Prospettive in prospettiva – Vedere, visualizzare, controllare). In: Prezioso, M. Capitale umano e valore aggiunto territorial. Geografia economica-politica/25. Aracne; 2018. p. 19-49

[24] Kostof, S. 1991. The City Shaped. Urban Patterns and Meanings Through History. London: Thames and Hudson; 1991. 352 p. ISBN 9580500280997

[25] Le Corbusier. Le Corbusier und Pierre Jeanneret: Ihr gesamtes Werk von 1929-1934. Zürich; H.Girsberger; 1935. 207 p.

[26] Watkin, D. Morality and Architecture: The Development of a Theme in Architectural History and Theory from the Gothic Revival to the Modern Movement. Chicago: University of Chicago Press;1984. 126 p. ISBN 0-226-87487-7

[27] Norberg-Schulz, C. Genius Loci: towards a phenomenology of architecture. London: Academy ed.; 1980. 213 p. ISBN 0856707007

[28] Tuan, Y.-F. Topophilia : a study of environmental perception, attitudes, and values. New York: Columbia University Press; 1974. 260 p. ISBN 0231073941

[29] Frampton, K. Modern Architecture: a critical history. London: Thames and Hudson; 1980. 324 p. ISBN 0500181748

[30] Enzenberger, H.M. Panopticon. London: Seagull Books Ltd; 2018. 144 p. ISBN 9780857425034

[31] Byggnads- och Brandstadgan för rikets städer. Stockholm: 1874 Byggnadsstadga och Brandstadga för rikets städer. 8 maj 1874 (boverket.se)

[32] Plan- och Bygglag, Stockholm: 1987:10. Available from: https://www.riksdagen.se/sv/dokument-lagar/dokument/svensk-forfattningssamling/plan--och-bygglag-198710_sfs-1987-10

[33] Bengs, C. The housing regime of Sweden: Concurrent challenges - Part A: Aims, effects and interpretations. Espoo: Aalto University publication series SCIENCE + TECHNOLOGY 14; 2015. 52 p. 978-952-60-6474-1 (electronic) Available from: The housing regime of Sweden: Concurrent challenges - Part A: Aims, effects and interpretations (aalto.fi)

[34] Wikipedia: Ambient intelligence. Available from: https://en.wikipedia.org/wiki/Ambient_intelligence read 3.1.2021

[35] ibid.

[36] Zuboff, S. The Age of Surveillance Capitalism. London: Profile Books; 2019. 691 p. ISBN 9781781256848

[37] Tegmark, M. Life 3.0, Being Human in the Age of Artificial Intelligence. London: Penguin Books Ltd; 2018. 364 p. ISBN 9780141981802

Chapter 9

# Internet of Things and Machine Learning Applications for Smart Precision Agriculture

*R. Sivakumar, B. Prabadevi, G. Velvizhi, S. Muthuraja, S. Kathiravan, M. Biswajita and A. Madhumathi*

## Abstract

Agriculture forms the major part of our Indian economy. In the current world, agriculture and irrigation are the essential and foremost sectors. It is a mandatory need to apply information and communication technology in our agricultural industries to aid agriculturalists and farmers to improve vice all stages of crop cultivation and post-harvest. It helps to enhance the country's G.D.P. Agriculture needs to be assisted by modern automation to produce the maximum yield. The recent development in technology has a significant impact on agriculture. The evolutions of Machine Learning (ML) and the Internet of Things (IoT) have supported researchers to implement this automation in agriculture to support farmers. ML allows farmers to improve yield make use of effective land utilisation, the fruitfulness of the soil, level of water, mineral insufficiencies control pest, trim development and horticulture. Application of remote sensors like temperature, humidity, soil moisture, water level sensors and pH value will provide an idea to on active farming, which will show accuracy as well as practical agriculture to deal with challenges in the field. This advancement could empower agricultural management systems to handle farm data in an orchestrated manner and increase the agribusiness by formulating effective strategies. This paper highlights contribute to an overview of the modern technologies deployed to agriculture and suggests an outline of the current and potential applications, and discusses the challenges and possible solutions and implementations. Besides, it elucidates the problems, specific potential solutions, and future directions for the agriculture sector using Machine Learning and the Internet of things.

**Keywords:** Machine Learning, Internet of Things, Agriculture, remote sensors, Land utilisation

## 1. Introduction

Précised agriculture depends on the utilisation of selective resources like water, fertilisers, seeds, and other necessary things. Sensor technology in the agriculture domain provides excellent support and offers the farmers to map their fields easily. Around the globe, the researchers of the agriculture domain strongly depending on the sensor technologies for both plant phenotyping and soil quality by using the latest technologies, including multispectral cameras, satellite imagery and drones, with the aid of internet of things (IoT) and cloud computing [1, 2]. The achievement of increment in the production level of agriculture outcome by introducing sensor technologies which offer the improvement in crop and soil quality, safety of food, sustainability, and profitability [2]. It helps farmers to understand the crops on the microscale. Sensors-based techniques used to provide appropriate tools to achieve the goals mentioned above [2]. Different sensing phenomena adopted for the agriculture field, and few of the selective sensors and their functionality.

### 1.1 Agriculture sensors

The technological advances and development facilities to attain the implementations on the agriculture domain by breaking the barriers to the basic needs of the farmers. Many sensing technologies that were already identified for precision agriculture by monitoring and optimising the crops [2]. Few of the sensors are listed below, which can offer the best solution for this precise farming.

#### *1.1.1 G.P.S. based position or location sensors*

This technology supports the proper application of agrochemicals and can safeguard water quality. Around 82 per cent of the implementation of the fertiliser can be uniform and appropriate by using a human resource controlled or lightbar guidance system [3]. Determination of longitude, altitude, and latitude by using the signals received from signals; these sensors can monitor the accurate position or location of the crop (**Figure 1**).

The G.P.S. systems used to measure the distances to the precisely located G.P.S. satellites to find positions on earth. Radio signals broadcasted from the G.P.S. satellites monitored by receivers [3]. A GPS position is usually determined by simultaneously measuring the distance to at least three satellites. The time taken for a radio signal which travels from the satellite to the G.P.S. receiver determines the length. For the calculation of positions, the information collected from the radio signals, which includes broadcasting time and satellite information, has to be processed.

This technology relatively inexpensive and also helps with parallel tracking devices, which assists the operators for the visualisation of the position concerning previous passes and to recognise the need for steering adjustments.

**Figure 1.**
*G.P.S. system.*

Commonly, these aids are coming with different configurations. G.P.S. technology was used for monitoring yield or mapping the field and also soil sampling [3, 4]. The G.P.S. navigation system can increase the efficiency of the farm and improve the aspects of agribusiness by reducing environmental impacts. This system can also reduce the operator's fatigue and anxiety regarding fertiliser and pesticide application. The use of this technology can demonstrate to the non-agricultural community that advanced technology used for farming efficiently and safely sampling [4].

### *1.1.2 IoT sensors*

In the last decades, farming implemented by several technological transformations and becoming more industrialised and driven by the latest technology. Introduction of smart agriculture gadgets which helps farmers for gaining best control on the process of crops growth and maintaining livestock as well with excellent efficiency. Internet of Things (IoT), based devices started to occupy every part of our life, from health care, automation, automotive and logistics, to smart cities and industrialisation (**Figure 2**). The Internet of Things creates up an era of precision agriculture sampling [5].

Precision agriculture is a basic term for all the services based on digital systems and inventions on technical things for the fulfilment of the modern farmer's needs for the yield optimisation, reduction of wastage, and maintaining the quality of environment [5, 6]. IoT sensors installed in the crop can support the farmers for allotting the pesticides and fertilisers in the right way along with the following support:

- Harvesting time optimisation
- The health of the crop
- Temperature, light and humidity level monitoring in greenhouses
- Soil quality and moisture level measurement

Many smartphone applications identified to incorporate with the Internet of Things (IoT) ideals, aggregation of data, and speed of the process, which may bring the data up to date, information can be provided to the small farmers like watering, seeding, fertilising and weeding. These applications are collecting the data from these sensors, especially from remote sensors and weather stations [6]. It helps in an in-depth analysis of data and provides valuable recommendations too.

Seeding is not guesswork after the innovation and application of IoT technologies. The programmed smart device can find the exact place for a seed to be planted and grown in a possible way. The collection of crops by the smart tractors with more exceptional efficiency and care when the harvest is ripe. Presently, the percentage of energy needed for the cultivation of crop by repairing the tractor damage itself goes around 80 to 90. By using the G.P.S. controlled steering system and route planning based on the input data, we can:

- Minimising erosion by tracking vehicle path
- Fuel cost reduction
- Improvement in accuracy on the operations

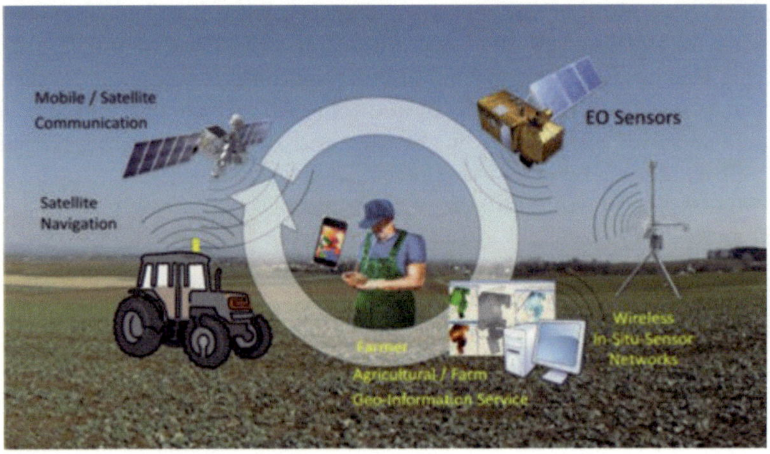


**Figure 2.**
*IoT system.*

The applications developed for small-scaled farmers may support them in multiple ways. The diagnosis of the diseases on plants identified and forwarded to the experts to rectify. The number of nutrients needed by the fertilisers by the determination of leaf colour and soil quality [7]. Also, the pH value of the soil and other conditions can be measured. From the observations on leaves, the water needs of the plants determined. The readiness on the crop harvesting with the aid of U.V. and white light-based photos can aid in the prevention of ripeness [7].

### *1.1.3 Optical sensors*

The optical sensors are used to collect and record the data about crop field and soil quality by the collection of light reflected from the growing plants. The application of nitrogen to the plants indicated to the users according to the health of the plants [8]. As this technology is not depending on the atmospheric light, the optical sensors used day and night. It uses external light to analyse the properties of soil. Measurement of light reflectance frequencies is carried out by the sensors in near and mid-infrared and polarised light spectrums. Optical sensors can be easily placed or integrated on vehicles or drones or even satellites too. The aggregation of data, collected from optical sensors, can be processed further. Determination of the organic matter, clay, and soil moisture level content can also be analysed by optical sensors (**Figure 3**).

According to the data collected using various platforms, like satellites, aerial (aeroplanes, UAVs and drones) and ground-based, the reflectance recorded. The collection of images from satellites, aircraft, and UAV's using cameras where the optical sensors installed in the ground are able to collect the reflectance data as a text file. According to the operation, these ground sensors classified either active or passive. The passive sensors are in need of an external source of light, like the sun. However, the active sensors are operated by their source of view of different wavelengths or a specific wavelength [9]. The relationship between the visible light and the chlorophyll content provides plant details. From this analysis, we could identify healthy plants as green. The mesophyll cells are reflecting the near-infrared light, which is invisible to the human eye, found that more than chlorophyll content, the quantity in a plant, results in the highest reflectance than the visible lights. Biomass production and evaluation of colour classified by analysing both wavelengths. Sensor position may affect the field measurements, like the crop distance, light

**Figure 3.**
*Optical system.*

source dependency, leaves may cover by snow dews, and also because of other factors that may cause the plant stress. The moderated distance between the target and the sensor kept avoiding noise in the captured signal. It will lead to overcoming the limitations of the sensor output. It is essential to monitor the leaves, which should not be covered by water molecules or dews, which may change the reflectance [9].

### *1.1.4 Electrochemical sensors and mechanical sensors*

Among different domains and their development like the Internet of Things (IoT) supported farming, the electrochemical sensor system is playing a vital role by detecting single or multiple soil components effectively, selectivity, and efficiently for soil quality measurements. It can be done either remotely by sharing the data and in-situ like the direct point of care on soil health. This perspective is aimed for the description of the state of art sensor technology based on the electrochemical mechanism for the measurement of soil quality by considering present scenarios. The electrochemical sensing mechanism explored its applications in many fields and even for a point of use. Mainly, lab-based methods like an ion-selective membrane, impedance spectroscopy, and amperometric sensors are in use to detect the nutrients of the soil and other parameters of agriculture (**Figure 4**) [10].

One of the attractive methods is to combine the electrochemical sensing technique by using ion-selective membrane transducers, which can easily monitor the parameters of soil like phosphate, nitrate, potassium, and others. Electrochemical sensing techniques are not so complicated like spectroscopy or any optical complexity and deployed directly to measure soil nutrients. These sensors are consisting of two electrodes of a working electrode, which can detect the target and another one of a reference electrode, which supplies a constant potential. The difference in potential between these two electrodes is either proportional or inversely proportional to the target according to its nature, either anions or cations. The working principle of this sensor governed by the Nernst equation. By relating the change in working electrode potential, which is compared with the potential of a reference electrode, based on the linearity of the activity of the sensed ion. The electrochemical sensors to deploy for in-situ measurements are expecting the electronic circuits embedded with the sensor (**Figure 5**) [11].

The microelectromechanical system (MEMS) based sensors embedded with electrochemical sensing units, which gains excellent potential for the analysis of soil quality because of their portability, rapidity, real-time measurement, and in-field deployability [12]. The ability of electrochemical soil sensors to sense different soil

**Figure 4.**
*Amperometric sensor.*

**Figure 5.**
*Electromechanical system.*

parameters, needed to be present in those systems as a basic and essential part for smart farming. This micro-scaled sensing system with the high potential for soil analysis is the much need for next-generation agriculture. MEMS-based sensors can save the data easily due to their affordability & sharing, on-time analysis, and accuracy in the decision [12].

### *1.1.5 Mechanical sensors*

These sensors used to estimate the mechanical resistance of the soil. The penetration or cutting through the land to measure the force using individual devices like strain gauges or load cells is the basic phenomenon of these sensors (**Figure 6**).

The developed prototypes by the researchers can map the soil resistance continuously in a feasible way. Unfortunately, these prototypes are not available commercially. A new technique called the "traction control" system on tractors based on drift sensors is using a similar method to control the three-point hitch on the way [13].

### *1.1.6 Dielectric sensors*

Dielectric sensors are used for measuring the soil moisture levels by the utilisation of the dielectric constant of the material. It defined as the electrical property, which is getting changed according to the content of soil moisture (**Figure 7**).

**Figure 6.**
*Mechanical sensor.*

**Figure 7.**
*Dielectric sensor.*

These sensors embedded with rain gauge stations and arranged around the farm. While the vegetation level goes down, the observation on soil moisture conditions can be performed by them. Also, the soil moisture sensors used the soil's dielectric constant to justify the content of the volume of water and the transmission of electricity based on the soil's capability depending on its dielectric constant. The dielectric constant land's water is larger compare with air, so that, if the water content of the soil increases, the increment of the dielectric constant of the soil will also be recorded. So, the constant dielectric measurement provides a fair observation of water content.

### *1.1.7 Airflow sensors*

Airflow sensors used to measure the permeability of air of the soil. The amount of pressure needed to pressurise a certain volume of air to some depth on the land, which is used to compare the multiple properties of soil (**Figure 8**).

From multiple experiments, it is possible to distinguish between various soil types and soil structure, moisture levels and compaction. These measurements can be made not only at a single location, while in motion too dynamically. The expected outcome is the need for pressure to allow a particular amount of air to the

**Figure 8.**
*Airflow sensor.*

ground in the wanted level of depth. By using such unique sensors, we can study various types of soil properties, including soil type, compaction, moisture level and structure, which produces unique identified signatures.

### 1.2 Benefits of agriculture sensors

Agriculture sensors can increase the food demand because of the utilisation of minimum resources like water, seeds, and fertilisers. These sensors fulfil the above basic requirements by resource conservation and field mapping. Also, these sensors easily installed and used efficiently. They are cost-effective too. Along with the usage in agriculture, these sensors can also serve for the prevention of pollution and global warming. With the advantages of communication protocols, these sensors controlled remotely.

### 1.3 Limitations of agriculture sensors

Precision agriculture and IoT technology are expecting flawless internet connectivity, which is a significant constraint and not available in many of the developing countries like I.N.D.I.A. there is a presumption among the customers that they may not be ready to utilise the present IoT devices integrated with agriculture sensors. Another significant impact on the infrastructure requirements like traffic systems, smart grids, and communication towers is not available everywhere, which also hinders the growth of the use of agriculture sensors.

Challenges and ideas to overcome limitations:

According to the expert's vision, precision agriculture has a standard potential to meet the increment in food demand around the globe. Even though the field has good growth and scope, still this has not robust as expected earlier. This domain has several challenges that we need to overcome.

a. The technology following the standards is not uniform and the same, which gets changed often. Precision agriculture expected, to a large extent.

The challenge depends on converting smart devices like sensors and gateways to farmer-friendly platforms.

b. Setting up the architecture for IoT technology is needed to be implemented. Knowledge of precision farming must be reached the farmers and enrich them to operate the sensors/tools independently so that the loss of the workforce prevented.

c. Providing continuous internet connectivity is mandatory, and network performance like the speed of bandwidth closely monitored.

d. All the crops are not going to produce the same products. So the product functioning must be defined correctly. Dividing their land as small zones for proper management may also derive the right results.

e. To prevent the mechanical damage of the sensor/device, continuous monitoring of the operation of these devices is a must. So, food safety cant is compromised. Upgradation of the tools is also essential. E-waste of these devices should adequately evacuate.

## 2. Soil quality identification for precision agriculture

One of the formidable global challenges is to feed the huge population soon. It predicted that the population could increases to 9.73 billion people by 2050 and estimated that it would require 70% additional food production in comparison to the present scenario [14]. The conventional agriculture practices resulted in a decline in the total productivity, causing poor ecological diversity, reduce the pollination services, affects carbon sequestration, causes soil and water pollution, soil erosion and food security [15, 16]. It is in dare need to use newly emerged modern sensing and controlling digital technology for effective agriculture. The agricultural sector is not just about maximising productivity it has shifted to the spectrum of other activities like optimising landscape management, development of rural, protection of the environment and social justice outcomes [17, 18]. Precision farming is one of the innovative methods practised, it incepted in the early 1980s, and with the past few years, it has become more common. It is a concept of "right practice at the right location at the right time at the right intensity". Precision agriculture uses electronic information and other digital technologies to collect data and analyse spatial/temporal data to improve the efficiency, productivity, and sustainability of agricultural operations [19]. Site-specific crop management practised from earlier decades like grid soil sampling and spot application of fertiliser and lime to optimise soil nutrient levels [20]. Global positioning systems (G.P.S.) initiated for civilian use in 1983, and in 1990's Global Navigation Satellite Systems (GNSS) enabled to develop equipment for variable rate fertiliser application for soil sampling and yield monitoring [21]. Incorporating digital management and surveillance technologies in farming automates the farming with integrated crop management to maximise the effectiveness of crop and yield [22–24]. The mechanical digitisation encompasses farm machinery for the sowing of seedling, fertilisers, cultivation, harvesting and the implication of satellites and tractors to drones, using Geographic Information Systems (G.I.S.), Global Positioning System includes yield mapping,

**Figure 9.**
*Precision farming cycle.*

remote sensing, variable rate irrigation, automatic tractor navigation, and robotics, proximal sensing of soils and crops, and profitability and adoption of precision farming (**Figure 9**). The details of the machinery discussed in the below sections. It is essential to understand the soil quality, functions and the role of indicators.

### 2.1 Soil quality

Soil is a vigorous component for crop production, and it plays a critical role in delivering ecosystem services. Like water and air, soils contribute a major carrier for biodiversity. The concept infers the capacity of soil to perform a specific function as a store, recycle and energy balance, that reflects the living and dynamic nature of the soil within the ecosystem boundary for multiple uses [25, 26]. The diverse potential of land uses to understand the quality of soil for ensuring the sustainability of the environment [27]. In the context of agriculture, good quality of soil has the fitness to support crop growth with enhanced productivity resulting in abundant and high quality of crops [28]. Generally, the soil has two parts viz., intrinsic, and dynamic. Intrinsic soils have the nature or inherent capacity for crop growth, which depends upon the parent material and topography. These soils are almost static, and the characteristics of these soils are permanent and do not change easily [29, 30]. Dynamic soil quality depends on its agronomic practices managed. The soil property encompasses soil texture, depth, permeability, soil organic matter, biological activity, water-and nutrient-holding capacity and soil structure. The organic matter changes from years to decades, pH changes from months to years, few properties can change from hours to days like microbial biomass and populations, soil respiration, nutrient mineralisation rates, and macroporosity [29, 31].

### 2.2 Soil functions

The primary function of soil is to nurture and sustain crop growth. Due to the dive's potential of land use, each soil performs a specific function for sufficient crop growth. Regulation of partition of water flow and storage helps for plant root penetration, and water infiltration for the crop growth [27, 32, 33]. The natural fertility of the soil increases by nutrient availability and has the adequate cation-exchange capacity, decreases acidity, maintains a proper buffer, and helps to remove the toxicants [34]. It also reduces the compaction risk like water retention, water infiltration, cohesion workability/trafficability [35–37]. The soil also reduces the contamination risk, leaching potential, toxic absorption, and toxic mobility. However, overuse exploitation of the earth can deteriorate the soil quality temporarily or permanently based on its usage. Soil erosion is widespread and estimated that approximately 75 billion tons of fertile soil is lost from world agricultural systems every year, consequently reduces the productivity of all-natural ecosystems [38–41]. Soil organic carbon (S.O.C.) observed and depleted 30–40% in cropland soils when compared to natural or semi-natural vegetation due to cultivation [42, 43]. Other threats like soil compaction, salinisation, waterlogging, nutrient imbalance, floods, and landslides and soil sealing, have both natural and human-induced causes [40, 41, 44–46]. This threat posses both agricultural production and terrestrial ecosystem. It reported that nearly 11.9–13.4% of the global agricultural supply lost due to soil degradation. Hence it is essential to protect soil degradation at different levels to safeguard food security, ecological health, and also for global sustainable development [47].

### 2.3 Soil indicators

Soil indicators fill the gap of traditional soil testing because merely measuring and reporting individual parameters is no longer sufficient; it requires an in-depth understanding of soil quality by inferring various parameters. U.S.D.A. classified the soil into four classes, such as visual, physical, chemical, and biological indicators. Visual was mostly observed to be a conventional type and mainly analysed by farmers through local knowledge and also obtained through photographic interpretation, subsoil exposure, erosion, presence of weeds and colour. The physical indicators connected to the organisation of the particles and pores like particle-size distribution, aggregate stability, max. Root depth, penetration resistance, hydraulic conductivity, infiltration rate, water holding capacity, water content, porosity, soil depth, particle density, water-dispersible clay, shear strength, stone content, clay mineralogy, total surface area, soil odour [48–56]. The chemical property such as pH; T.O.C. or organic matter, Nutrient Availability electrical conductivity; selected heavy metals, organic pollutants, particulate matter [55–66] Soil respiration; N. mineralisation, earthworms, nematodes respiration, urease activity enzyme activities, total species number, fungal biomass functional diversity, bacterial biomass, potential denitrification activity, potential amonium oxidation, mycorrhiza populations root health, soil fauna diversity, phosphatase activity, microbial diversity are the biological indicators that measure the quality of soil [49, 54, 67–74]. The selection of these indicators needs to ensure that they are sensitive and responsive to pressure and change in land use management. **Table 1** infers that indicators measured for different countries (**Table 1**). Soil indicators refer to the capacity of soil to perform crop production that used in response to the dynamic changes in an agroecosystem.

| Indicators | Values | Description | References |
|---|---|---|---|
| pH ($CaCl_2$) | 4.0 ± 0.37 | Physical and chemical properties of soil in Araucaria forest (N.F.), Brazil | Pereira et al. [75] |
| Organic-C (g $kg^{-1}$) | 33 ± 12.9 | | |
| Bulk density (g $cm^{-3}$) | 1.08 ± 0.2 | | |
| Macroporosity ($m^3$ $m^{-3}$) | 0.16 ± 0.07 | | |
| Microporosity ($m^3$ $m^{-3}$) | 0.41 ± 0.06 | | |
| Sand (g $kg^{-1}$) | 459.0 ± 157 | | |
| Silt (g $kg^{-1}$) | 87.3 ± 40 | | |
| Clay (g $kg^{-1}$) | 453.8 ± 136.5 | | |
| Organic matter | 10–20 g $kg^{-1}$ | | Lal [76, 77] |
| Nitrogen | 1.6–2.4 g $kg^{-1}$ | | Adeoye and Agboola [78] |
| Active carbon | 6–15 g $kg^{-1}$ | | Adeyolanu [79] |
| Cation exchange capacity | 3.5–6.0 c mol $kg^{-1}$ | | Adeoye and Agboola [78] |
| Wet stable aggregate | 0.40–0.75 kg $kg^{-1}$ | | Adeyolanu [79] |
| Mean weight diameter | 0.53–2.00 mm | | Adeyolanu [79] |
| Available moisture content | 8–20% | | Lal [76, 77] |
| Bulk density | 1.3–1.5 g $cm^{-3}$ | | Lal [76, 77] |
| Macroporosity | 0.15–0.18 m3 $m^{-3}$ | | Lal [76, 77] |
| Soil strength | 60–120 kPa | | Adeyolanu [79] |
| Infiltration capacity | 7–21 cm $hr.^{-1}$ | | Adeyolanu [79] |
| Saturated hydraulic conductivity | 0.2–3 cm $hr.^{-1}$ | | Adeyolanu [79] |
| Organic matter content(%) | 4.3 | Benchmark soil, for natural Pampa Region, Argentina | de la Rosa and Sobral |
| Respiration rate (kg C $ha^{-1}$ $d^{-1}$) | 83 | | |
| Aggregate stability (%) | 70 | | |
| Infiltration (mm $h^{-1}$) | 44 | | |
| Compaction (Mpa) | 3.7 | | |
| O.M. (%) | 2.65 ± 0.96 | Soil water retention and soil resistance to penetration curves of Argentina | Imhoff et al. |
| Clay (%) | 27 ± 10 | | |
| Sand (%) | 18 ± 18 | | |
| Silt (%) | 55 ± 15 | | |
| Bd (g $cm^{-3}$) | 1.37 ± 0.09 | | |
| B.D. (g $cm^{-3}$) | 1.5 | Soil quality indicators, baseline limits used for in northern Ethiopia. | Harris et al. [80] |
| MWHC (%) | 30 | | Gregory et al. [81] |
| $OC^e$ (%) | 3.5 | | Kay and Anger [82] |
| SAS (%) | 30 | | Harris et al. [80] |
| Zn (mg $kg^{-1}$) | 18 | | Mausbach and Seybold [83] |
| Fe (mg $kg^{-1}$) | 40 | | Harris et al. [80] |

**Table 1.**
*Different types of indicators used for different countries.*

## 3. Comprehensive machine learning models in agriculture

Ml is a technology that aims to build an intelligent model that makes an accurate prediction without the intervention of human beings. The conventional machine learning approach depicted in **Figure 1**. It constructs various algorithms to make effective decisions in the problem domain. The primary step is to select the data on the problem under investigation and to select the parameters for the examination. The model is trained by a sample set of data (termed as training data) to gain experience in the environment and make the model fit. Later, the model evaluated using a sample set of data (termed as test data). So this is the primary step involved in any machine learning model, i.e., Train-Test-Predict. Usually, the data set was divided into two viz., training (70%) and testing (30%). Testing data is kept separate and not used in the preparation. The conventional machine learning approach depicted in **Figure 9**.

The dataset with many alternatives is collected and pre-processed using any normalisation or standardisation methods. The pre-processed data set was divided as train and test data set. The machine algorithms take the train data as input to train the model or to learn for the historical information. The trained model is evaluated with test data. The data visualisation tools are used for visualising the prediction or classification results. Algorithms involved in machine learning are supervised and unsupervised learning. In supervised learning, the model is trained with input data and mapped it into the known results whereas, in unsupervised learning, the model is trained, validated with input data and finds all type of unknown patterns.

The most familiar learning models that fall under these two categories are clustering, regression, classification, and dimensionality reduction. Machine learning utilises a secondary dataset (termed as validation data) for training the model further to avoid the overfitting of the model by the trained data. If the model generates more error on validation data, that means the model overfitted with the prepared data so that training stopped. Now the data split can be done like 60, 10, and 30 per cent of training, validation, and testing, respectively. Machine learning employed in almost all scientific applications such as health care, home automation, smart city, robotics, aquaculture, digital marketing, financial solutions, enterprises, climatology, food safety, agriculture, and more.

As Agriculture forms the major economy for most of the countries, better assistance speeding up each stage of agricultural crop production is mandatory. ML and the Internet of Things (IoT) serve this platform more effectively. IoT devices such as sensors, actuators through wireless communication protocols continuously monitor the crop, soil, water and communicate their health to remote devices either by message or log data or buzzer to alert the agriculturalist to take necessary actions. The data from these devices will make meaningful predictions and recommendations to the user exclusively farmers through machine learning algorithms.

Machine learning models trained by the historical data of the agricultural field through which it gains experience and makes wise decisions for the data signals received from the IoT devices. The data collected from these IoT devices must be secured and ensure confidentiality for accurate prediction results. Precision Agriculture is a strategy adopted to integrate heterogeneous information (Spatio-temporal data) for making precise and effective managerial decisions for global sustainable agricultural practices. Most of the parts of our country are adopting this strategy to improvise agrarian production in a brief span. Application of machine learning in precision agriculture has reshaped the plan such as field-based crop suggestion, fertiliser recommendation, water supply prediction, harvest prediction, thereby controlling the water usage by assisting the agriculturalists or farmers for better yield in a smart way.

Digital agriculture (a term coined by use of Precision Agriculture and Remote sensing) evolved to increase agricultural productivity with a minimised impact on environmental factors. Digital agriculture uses the data (crop, soil, and weather) sensed from the IoT devices to make effective decisions on nutrient demand-based fertiliser recommendation, water supply through proper irrigation, soil nourishment, pest or weed control, and crop protection from intruders. Digital agriculture focuses on the best-of-breed optimisation algorithms fro crop production and its protection during growth. Multi-cropping is a technique adopted in Digital agriculture or smart farming, which allows the cultivation of more than one crop in a single cultivable land.

Digital agriculture has to take more precautionary steps while feeding these different crops with weeds and fertilisers as the mixed plant has a different nutritional requirement and water supply. So it takes into account inter-variability and intra-variability among the crops before feeding the fertilisers. It adopts the techniques like in-row treatment to spray fertiliser for each plant separately, sensor-equipped drones to track the weed, automated sensing of fertiliser details from the barcode label for a correct proportionate mix of pesticides, drift reduction techniques and integration of these applications with global positioning system and comprehensive information system for periodic relay to the agriculturalists.

The application of Machine learning in different stages of agricultural crop production are depicted in **Figure 10**. The necessary steps involved in crop cultivation are Land suitability analysis, appropriate crop selection, crop production, crop protection, nutrient supply, water supply, crop health monitoring (pest and weed control), human and animal attack detection, yield management, and post-harvesting.

Although these steps are common for all types of crops, soil nourishment value and chemical composition determine the techniques adopted in each level. Also,

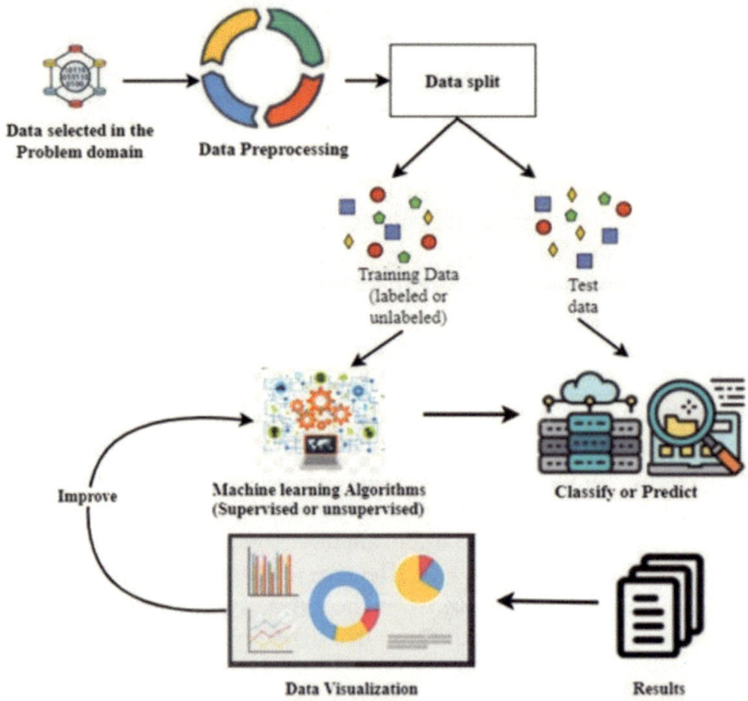


**Figure 10.**
*Machine learning approach.*

this paves a significant consideration of fertiliser supply when multi-cropping is selected. This multi-cropping technique has been in evolution decades back and done explicitly in the hill areas with meagre farming areas yielding better productivity.

### 3.1 Machine learning in land suitability analysis

Land suitability analysis has done for any barren land before permitting any residential plots to be constructed on that land. By ensuring better land use analysis, most of the agricultural land not converted into residential buildings or industrial areas. It will reduce the cultivable land area and air pollution. Cultivating a crop without suitability analysis may lead to an enormous waste of time, more fertiliser supply, abnormal and water requirements. Therefore, Land suitability analysis for the cultivation of crops is an essential factor in ensuring sustainable agriculture yielding better production. Geographic Information System (G.I.S.) provides more significant support in aiding the suitability analysis of the land. Multiple factors considered for analysing the land suitability attained from advanced G.I.S. systems. Some of the factors considered for land suitability analysis are soil quality parameters (pH, organic carbon content, salinity, texture, slope), topography, water availability, essential nutrients, socio-economic factors.

Mokkaram et al. have implemented an ensemble classifier method, namely RotBoost, an integration of Rotation forest and AdaBoost algorithms for land suitability analysis. Benjamin et al. have assessed the suitability of land for cultivation of a different variety of rice crops in rural Thailand using species presence only prediction method. They proved that the MaxEnt model outperforms and provides better crop suitability on particular land. A land with a higher suitability index for the cultivation of a crop selected for farming. Support Vector Machines (SVM) preferred for classifying the suitable area for agriculture of rainfed wheat based on thirteen factors relating to property, topography, climate, and soil.

Senagi et al. have applied Parallel Random Forest (PRF), SVM, Linear Regression (L.R.), K.N.N., Linear Discriminant Analysis (LDA), and Gaussian-Naïve Bayesian to ensure the land suitability for sorghum crop cultivation. PRF provides better accuracy than others when evaluated using ten cross-fold validation. One of the most important attributes that contribute to suitability analysis is soil quality. The moisture content in the soil helps to determine the suitability of growing a particular crop in a land. Typically the dryness or wetness level of the earth can be determined by considering the same at other locations, which has similar soil type and hydroclimate.

Coopersmith et al. recommend that land suitability analysis will be more accurate in the sandier soil (with more drainage) than poorly drained soils. They have used K.N.N., Boosted perceptron, and classification tree for soil dryness estimate at a site in Urbana. Perhaps, K.N.N. shows best results than Boosted Perceptron when evaluated with farmer's assessments. Soil fertility levels should be periodically monitored and maintained at appropriate levels for the continuous nourishment of crop production in agricultural land. Gholab applied the decision tree classification model for building the predictive model. All these approaches use the data obtained through remote sensing and IoT devices. A better understanding of the land suitability of the agricultural field under consideration will assist in selecting suitable crops as well as supplying fertiliser to make it better nourished for growing the required plants. It followed by crop production, water supply, and Nutrient management.

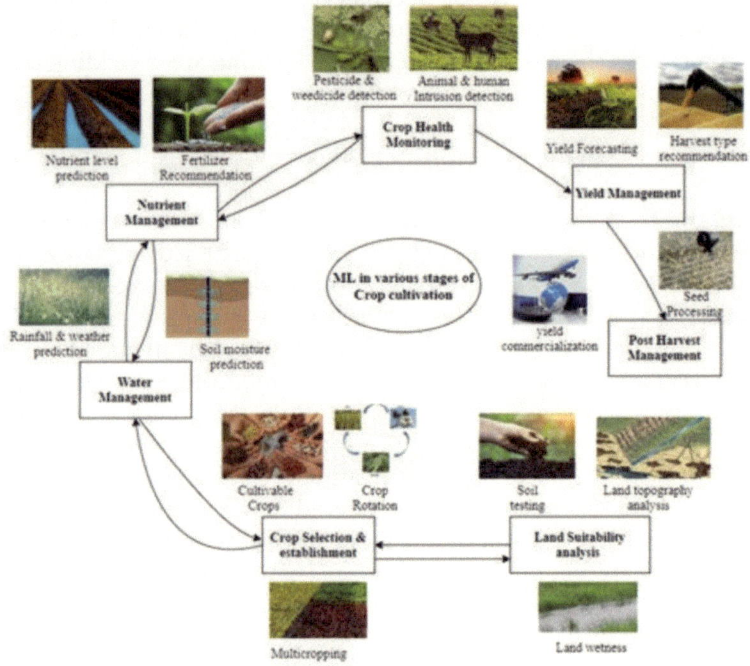


**Figure 11.**
*Machine learning in agricultural crop cultivation.*

## 3.2 Machine learning in crop production

Crop Production and management include crop selection, soil preparation based on suitability analysis, sowing seeds, application of manure & fertiliser, water management through proper irrigation mechanisms, and harvesting. Machine learning in agriculture crop production links various participants in the food chain or agricultural chain. Machine learning helps the agriculturalists in making better decisions in crop quality determination, yield prediction, plant species determination, crop disease prediction, and harvesting techniques (**Figure 11**).

The machine learning algorithms data acquired from IoT sensors in the agricultural field. Once the data feed, ML algorithms train the model using history and can make predictions at any stage of production to determine the different features required to predict the yield. It will help to improve the nutritional value (if deficient in the current return predicted) in the next production. Consequently, the crop production price will show a dramatic improvement in the upcoming yield. Application of A.I. in agriculture will enable the farmers to get up to minute information about current production, suggestions on next production, plant species identification, and quality improvement.

Once Land suitability analysis for cultivation is done, crop species selection has to be done based on suitability. Based on the nourishing factor in the soil and nutrient capability, a crop can be selected appropriately. Multi-criteria decision-making models used to get land suitability analysis. Image processing techniques integrated with machine learning suggested for plant species identification for the given crop image. Patil et al. analysed the various ML techniques used for crop selection based on environmental parameters and live market. They have used the K.N.N. classifier for the data obtained through multiple IoT sensors and prices based on entries in National Commodity and Derivative Exchange.

Land specific yield prediction by considering Crop yield prediction using topological algorithms like ANN, backpropagation, and Multi-layered perceptron through the implementation of a neural network. Support vector regression (S.V.R.) a variant of SVM used for crop yield prediction. As nitrogen is an essential component for photosynthesis, nitrogen management is mandatory as the yield prediction. The various decision support systems provide agricultural decisions, the agriculturalist has to deal with enormous heterogeneous data for making wise decisions, so Machine learning plays a vital role. Chlingaryan et al., 2018 have analysed the various ML approaches and signal processing methods used for crop yield prediction and optimised techniques for nitrogen management. They reviewed that B.P.N.N. provides best accurate crop yield estimation (by considering the importance of vegetative indices), CNN with Gaussian Process is best for feature extraction, best Multi-class crop estimation by M5 Prime R.T., Least Squares SVM for Nitrogen management and Fuzzy cognitive map for representing the expert's opinion.

A comparative analysis of ML algorithms M5-Prime, K.N.N., S.V.R., ANN, and Multi-linear regression model was carried out on prediction of crop yield and suggested that M5-Prime outperforms others followed by K.N.N., S.V.R., ANN, and the last Multi-Linear Regression. It was evaluated based on the accuracy metrics (Normalised Mean Absolute Error, Root Relative Squared Error, Root mean square error, and Correlation Factor). Corn yield prediction predicted by Back Propagation Neural Network whose efficiency tested on green vegetation index, Normalised Difference Vegetation Index, perpendicular vegetation index, and soil adjusted vegetation index. Also, Deepa learning showed the most stable results on corn yield prediction at the particular region (Iowa state) when compared with Estimated Randomised Trees, Random Forest, and SVM. Deepa learning overcomes the overfitting problem prevalent in most of the ML algorithms.

One or more stages of crop cultivation will give information to other steps and vice versa. Depending on soil test results done during land suitability and crop health monitoring, the fertilisers will be recommended. Consequently, water and nutrient management carried out. The ML approaches work best for fertiliser recommendation. Water management is M.L.P. neural network with Backpropagation algorithm based on soil nutrient content, Gradient boosting and Random forest for soil nutrient assessment and Multivariate Relevance Vector Machine and Multilayer Perceptron for estimating the water requirement based on evapotranspiration and climatic data. Periodic Drought assessment is essential for crop maintenance and water management. Machine learning approaches used for drought assessment are Random Forest, Cubist, boosted regression trees, support vector regression, coupled wavelet ANNs, and ANN. Drought assessment is done based on the drought factors (land surface-related) and drought index.

### 3.3 Machine learning in crop protection

Crop protection implies the protection of crops from weeds (unwanted plants that grow in the land), pests (insects, bugs), and intruders (an animal which intends to graze the crops and human for theft). K-Means clustering, Support vector machines, and Neural networks are more prominent machine learning techniques employed in Precision Agriculture for crop protection. The weeds may cause a significant loss to the crop yield. Weedicides are applied (weeding) before the crop seeding stage and flowering stage. The weedicides, instead of any common herbicide, have to be explicitly asked to avoid the devastation of the desirable crop in the field. Accurate detection of weeds is more significant and done using Machine learning algorithms integrated with sensor data.

One of the most undesirable weeds, which causes a significant loss to crop and very difficult to detect and abolish, is Silabynum marianum. Pantazi XE et al., have suggested a weed detection method by multispectral imagery obtained through a camera mounted on Unmanned Aircraft Vehicle (UAV) using Counter Propagation ANN, XY-Fusion Network and Supervised Kohonen Network (S.F.N.) to detect Silabynum marianum from other crops. Furthermore, a weed detection system that accurately classifies the weeds was designed based on hyperspectral images through the camera mounted on a robot using an active learning machine learning model. This model designed using a class neural network classifier (one class mixture of Gaussians) for novelty detection and one self-organising class map. This active learning model provides 100% accuracy on the classification of the crop, whereas different weed species detection accuracy varied from 34 to 98%.

The different weed species detected using this model are *Taraxacum officinale*, *Ranunculus repens*, *Poa annua*, *Cirsium arvense*, *Stellaria media*, *Urtica dioica*, *Sinapis arvensis*, *Oxalis europaea*, *Polygonum persicaria*, and *Medicago lupulina*. The model outperformed when compared with the autoencoder network and one-class SVM classifiers. Some of the other weeds detected through images from cameras on UAV using machine learning techniques are: identification of broadleaf and grass from soybean using Convolution Neural Network (CNN) in comparison with SVM, Random Forest and Adaboost; weeds classified in sugarbeet fields with sugarbeet shape features using SVM and ANN are Pigweed, Turnip weed, Lambsquarters and Hare' s-ear mustard.

Some pests may infect weeds, and that might be contagious to the crops, so pest detection is one of the essential stages in crop protection. Thus weeds serve as hosts for pests and diseases consuming all the resources supplied to the plants. It is done using machine learning algorithms and followed by the recommendation of pesticides for pests. The images acquired through the optical sensors attached to UAV help in detecting the pests. CNN provides better results in this classification of pests from images. D. C. Corrales et al. have suggested a list of supervised machine learning algorithms used for crop protection in terms of diseases and pests. The are SVM, K.N.N., ANN, Decision trees, and Bayesian Network. Decision trees, SVN, and ANN, are best for prediction and classification of pests, whereas Bayesian Networks and K.N.N. are excellent in training. These pests have a devastating effect on the crop storage, precautionary measures taken by identifying the categories of pests and their nature of the occurrence. Crop Image analysis used to categorise the type of pests using computer vision.

Cheng et al., have implemented a deep residual learning model for classifying the pest image and it outperforms the Back-Propagation Neural Network and SVM in the accuracy of the pest image recognition. Also, it provides better performance than deep CNN (Alexnet). Tomato Whitefly classification using deep CNN, Paddy crop pests classification using deep CNN [84] and banana pest and disease detection using deep CNN are some of the successful CNN based crop pest classification models outperforming the traditional approaches. Therefore integration Image processing or computer vision and machine learning CNN algorithms provide the best classification of crop pests and diseases.

Animal intrusion detection is one of the threats to the agricultural crop. These intrusions identified and detected to avoid loss of crop production. IoT sensors provide periodic alerts on the detection of an animal object like rats, cow, sheep, elephant, and other wild animals. It can be detected effectively and prevented through wireless sensors alerts to farmers mobile and machine learning algorithms can be used for object classification. Also, Machine learning algorithms used to predict the animal or human object entry apriori by training the model with past data from IoT sensors.

### 3.4 Machine learning in livestock management

Livestock management is essential for animal husbandry, and wellbeing of rural people as this frames a significant economic factor for rural beings and sustainable agricultural practices. Livestock species used for varied purposes such as employment for the community, food supply, nourishing the family nutrition, significant income to few families, soil enrichment, believed ritual events. Livestock management includes vaccination for cattle species, health monitoring of livestock, managing livestock during drought, feed schedule, grazing, milk quality management, ketosis for dairy animals, ear tagging, production, and castration. The machine learning approaches used for animal welfare are Bagging with decision trees for classification of cattle behaviour-based features like grazing, walking, sleeping, ruminating, classification of chewing patterns in calf using decision tree/C4.5 based on chewing signals while dieting ryegrass, supplements, hay, rumination and during sleep, behavioural changes monitoring and tracking of pigs using Gaussian Mixture Model based on 3D motion information, ANN for determination of rumen fermentation, CNN for face recognition of pigs, estimation of beef's carcass weight using S.V.R. models, SVM models for early evaluation of egg production in hen and bovine weight estimation in cattle.

### 3.5 Considered machine learning approaches for agriculture

Several machine learning approaches have become popular for achieving superior and precision agriculture [85, 86]. The following sub-section discusses certain machine learning approaches that have been deployed for achieving enhanced agricultural benefits. In the perspective of machine learning, supervised learning is a phenomenon that encompasses both the input and the sought after target values. Besides, both the input and target data are in labelled form, which offers a learning platform for processing data in the future. Further, when this model is offered a new test dataset (with a similar background) since the model is already trained, it generates the accurate output for the test data. Kaur et al. review the scheme of plant disease diagnosis and taxonomy employing leaf images with the aid of computer vision technologies [87].

#### *3.5.1 Belief Networks*

Belief Networks also referred to as Bayesian Networks, are probabilistic graphical models, which are utilised for building models from data or through specialists' outlook. Further, these networks can be a beneficial approach for evaluation and effective decision-making process in the case of agrarian problems. The Belief Networks are built using the Bayes theorem, which in turn supports in computing the input data's posterior likelihood. Belief Networks are more suitable for agrarian applications owing to their capability to reason with inadequate data, and further, they also add new evidence data. Further, Aguilera et al., [88] evaluate the quality of the groundwater by deploying the probabilistic clustering supported by the hybrid Bayesian networks via Mixtures of Truncated Exponentials. Huang et al. [89] established a Bayesian driven averaging technique for offering a trustworthy forecast of maise yields in China. Besides, Cornet et al. [90] established a Bayesian network model for identifying the initial growth and yam yield interactions.

Zhu et al. [91] established the Bayesian networks based model to characterise the connections between the symptoms and harvest maladies. De Rainville et al. [92] devised the naive-Bayesian classifier combined with the Gaussian mixture clustering approach for classifying the weeds from the actual row crops. Stanaway et al. [93] discussed the hierarchical Bayesian framework for the early diagnosis of exotic plant pests attacks and infectious plant diseases. Russo et al. [94] established a Bayesian model for estimating the hydrologic characteristics and irrigation needs in order to devise a sustainable water management scheme for the agrarian lands in Punjab, India.

### *3.5.2 Classification and regression trees*

The classification and regression trees (CART) are usually referred to as decision trees. Besides, they act as a decision support tool, which deploys a tree-like graph or a decision model and their probable consequences. In a decision tree, each internal node signifies a test on a feature, each branch characterises a result of the test, and each terminal node embraces a class label. There are several applications of the decision tree in agriculture, such as disease diagnosis and classification, crop monitoring and weed classification. Waheed et al. [95] devised a CART algorithm for categorising hyper-spectral information of the corn plots into different classes based on water stress, weeds' existence, and nitrogen application rates. Xueli Liu et al. [96] established a decision tree model for assessing grain loss due to various factors involved in grain storage. Bosma et al. [97] discussed the decision tree model for estimating and modelling the decision-making process of the agriculturists on assimilating aquaculture into agronomy in Vietnam. Moonjun et al. [98] concerted on deploying the G.I.S. assisted decision tree and artificial neural network-based model for assessing the landscape-soil relationship in inaccessible areas of Thailand. Kim et al. [99] established the decision-tree assisted model combined with the geographical information system for forecasting and mapping the variety of bacteria in the soil. Rossi Neto et al. [100] elucidated a decision tree-based approach for categorising the biometric attributes with the highest impact on the sugarcane productivity under the distinct arrangement of plants and edaphoclimatic settings.

### *3.5.3 Connectionist systems*

Connectionist systems also referred to as an artificial neuron network (ANN) is a computation based archetypal relying on the structure and functions of the human brain. Moreover, the connectionist systems are known to possess the neurons that are interconnected to one another in numerous layers of the networks. Also, such neurons are referred to as nodes. Connectionist systems consist of input and output layers, as well as a hidden layer comprising of units, which converts the input into unique values that the output layer can use. Besides, such systems are exceptional methods for determining complicated patterns. Also, brain-inspired systems have an arithmetical value that can accomplish more than one task, concurrently. Priyanka et al. [101] discussed the deployment of the neural networks combined with satellite imageries for monitoring crops and also for estimating the agricultural produce. Daniel et al. [102] established a review on ANN modelling for Agroecology application. Jha et al. [103] investigated various the usage of ANN/Artificial intelligence techniques combined with the internet of things and wireless systems for classifying plants and flowers, in order to accomplish sustainable development in the agricultural domain. Kaul et al. [104] deliberated about the deployment of the ANN models for forecasting the corn and soybean produces under distinctive climatic settings in Maryland, U.S.A. Thomas et al. deployed the multilayer neural networks along with genetic algorithms for detecting the viruses in plants via data collected using biosensors. Were et al. [105] employed the ANN approach for forecasting and mapping soil organic carbon stocks in Kenya. Besides, this model was validated by means of independent testing data. Nahvi et al. [106] deployed a self-adaptive evolutionary model for forecasting the everyday temperatures of the soil, at six diverse depths and validated the results through genetic programming and ANN models.

### *3.5.4 Random forest*

Random forests (R.F.) algorithm is a supervised learning approach that is deployed for real-world or simulated applications (both classification and regression problems). Besides, it is similar to the bootstrapping algorithm combined with the CART model. Moreover, in this algorithm, the decision trees on data samples get created, followed by the forecast from each of these trees, and lastly, chooses the best solution via voting. Further, it is an ensemble technique that performs superior to a solitary decision tree, since it lessens the over-fitting by averaging the outcome. Fukuda et al. [107] devised an R.F. model for forecasting the yield of the mangoes in retort to the supply of the water in diverse irrigation systems. Philibert et al. [108] designed an R.F. model for forecasting the N2O discharge depending on local data for ranking environmental and crop management attributes. Further, they also established the impact of these attributes on N2O emission. Rhee et al. [109] elucidated an RF-based high-resolution drought estimation system for ungauged expenses by deploying the long-range climate estimation and remote sensing information. Inacio et al. [110] developed a system for identifying weeds in sugarcane fields by deploying the Unmanned Aerial Vehicle for capturing images and later classifying these images via an RF-based classification scheme. Saussure et al. [111] demonstrated the harms caused in maise crops due to wireworms in several agricultural fields across France. Besides, they deployed the R.F. technique for imputing the missing values. Everingham et al. [112] devised an R.F. model for categorising the different types of sugarcane and crop cycle with the aid of imagery acquired via hyperspectral sensors.

### *3.5.5 Support vector machine approach*

A support vector machine (SVM) is a comprehensive supervised learning approach, which is generally deployed for mostly solving two-class categorisation problems. Besides, the SVM can also be utilised for analysing the data for classification and regression scenarios. Further, SVM employs the kernel phenomenon for transforming the data and then depending upon these transformations; it determines an optimal borderline among the likely outcomes. Moreover, the decision boundary between the two classes on a graph needs to be widespread. SVM builds an optimal borderline that splits the new data point and assigns it to the correct category. Therefore, this optimal borderline is also known as the hyperplane. Misra et al. [113] investigated the deployment of SVM techniques for stimulating run-off and sediment produces from the watersheds, via the support of the monsoon-period information. Kovačević et al. [114] developed an SVM model for classifying soil types based on the assessment of the physical and chemical characteristics of the soil. Huang et al. [115] devised a machine vision-driven SVM system for diagnosing the borer diseases in the sugarcane plant. Kawamura et al. [116] devised an SVM model for classifying the diverse inflorescence types by making use of an artificial dataset. Liu et al. [84] developed an SVM-based system for classifying the urban soil based on quality attributes, such as the soil toxicity due to heavy-metals, soil richness, and potency. Singh et al. [11] reviewed the deployment of SVM based model for the assessment of the plants undergoing high-throughput stress phenol-typing, with the aid of sensors.

## 4. Conclusions

In this chapter, smart sensor-based approaches are presented for precision agriculture. The use of remote sensors like temperature, humidity, soil moisture, water level sensors and pH value, will provide an idea to on productive farming, which will show accuracy as well as practical agriculture to deal with challenges in the field. This advancement could empower agricultural management systems to handle farm data in an orchestrated manner and increase the agribusiness by formulating effective strategies. The evolutions of Machine Learning (ML) and the Internet of Things (IoT) established methods offered to help researchers to implement these methods in agriculture to support farmers. These will support farmers to improve throughput, effective utilisation of field and manage pests. This paper presents to contribute to an overview of the modern sensor technologies deployed to precision agriculture and suggests an abstract of the present and essential applications and presents the challenges and feasible solutions and applications.

## Author details

R. Sivakumar[1*], B. Prabadevi[2], G. Velvizhi[3], S. Muthuraja[1], S. Kathiravan[2], M. Biswajita[4] and A. Madhumathi[5]

1 School of Electronics Engineering, Vellore Institute of Technology (VIT), Vellore, India

2 School of Information Technology and Engineering, Vellore Institute of Technology (VIT), Vellore, India

3 $CO_2$ Research and Green Technologies Centre, Vellore Institute of Technology (VIT), Vellore, India

4 School of Agricultural Innovations and Advanced Learning, Vellore Institute of Technology (VIT), Vellore, India

5 School of Planning and Architecture, Vellore Institute of Technology (VIT), Vellore, India

*Address all correspondence to: rsivakumar@vit.ac.in

## References

[1] Chuli Hu; Xiang Zhong; Jingyue Xu (2014), 'Study on integrated discovery system of sensors for agriculture observation application,' The Third International Conference on Agro-Geoinformatics Agro-geoinformatics, IEEE Digital library.

[2] Gonzalo Pajares, Andrea Peruzzi, and Pablo Gonzalez-de-Santos (2013),' Sensors in Agriculture and Forestry', Sensors-Open access journal, 13(9), 12132-12139.

[3] Castrignanò, A.; Wong, M.T.F.; Stelluti, M.; De Benedetto, D.; Sollitto, D. In *Geoderma (2012),* 'Use of E.M.I., gamma-ray emission and G.P.S. height as multi-sensor data for soil characterisation,' Geoderma, (175-176), 78-89.

[4] Fountas, S.; Paraforos, D.; Cavalaris, C.; Karamoutis, C.; Gemtos, T.A.; Tagarakis, A.; Abu-Khalaf, N (2013), 'A five-point penetrometer with G.P.S. for measuring soil compaction variability,' Computers and Electronics in Agriculture, (96), 109-116.

[5] Chun-Ting, Pan; Meng-Ju, Lee; Nen-Fu, Huang; Jhong-Ting, Luo; Jia-Jung, Shao (2020), 'Agriculture Blockchain Service Platform for Farm-to-Fork Traceability with IoT Sensors,' *International Conference on Information Networking (I.C.O.I.N.),* 158-163.

[6] Dagar, Rahul; Som, Subhranil; Khatri, Sunil Kumar (2018), 'Smart Farming – IoT in Agriculture,' *International Conference on Inventive Research in Computing Applications (I.C.I.R.C.A.),* 1052-1056.

[7] Sushanth, G.; Sujatha, S (2018), 'IoT based smart agriculture system,' *International Conference on Wireless Communications, Signal Processing and Networking (WiSPNET),* IEEE Xplore Digital Library.

[8] Lee See Goh, Norikazu Kumekawa, Kazuhiro Watanabe, Norihiko Shinomiya (2014), 'Hetero-core spliced optical fiber S.P.R. sensor system for soil gravity water monitoring in agricultural environments,' Computers and Electronics in Agriculture (101), 110-117.

[9] Nivedita Singh, Aditya Narayan Singh (2020), 'Odysseys of agriculture sensors: Current challenges and forthcoming prospects,' Computers and Electronics in Agriculture, (171), 105328.

[10] Srivastava, Anup K.; Dev, Atul; Karmakar, Surajit (2018), 'Nanosensors and nano biosensors in food and agriculture,' Environmental Chemistry Letters, *(16), 161-182.*

[11] Arti Singh, Baskar Ganapathysubramanian, Asheesh Kumar Singh, Soumik Sarkar, Machine Learning for High-Throughput Stress Phenotyping in Plants, Trends in Plant Science, Volume 21, Issue 2, 2016, Pages 110-124, ISSN 1360-1385, https://doi.org/10.1016/j.tplants.2015.10.015.

[12] Tyronese Jackson, Katrina Mansfield, Mohamed Saafi, Tommy Colman, Peter Romine (2008), 'Measuring soil temperature and moisture using wireless MEMS sensors', measurement, (41), 381-390.

[13] Qingyuan Zhu, Wei Chen, Huosheng Hu, Xiongfeng Wu, Chunsheng Xiao, Xinyi Song (2019), 'Multi-sensor based attitude prediction for agricultural vehicles,' Computers and electronics in agriculture, (156), 24-32.

[14] F.A.O., "The future of food and agriculture - trends and challenges," 2017, http://www.fao.org/3/a-i6583e.pdf. View at Google Scholar

[15] Tilman, D., Balzer, C., Hill, J., Befort, B.L., 2011. Global food demand and the sustainable intensification of agriculture. Proc. Nat. Acad. Sci. 108 (50), 20260-20264.

[16] Tscharntke, T., Clough, Y., Wanger, T.C., Jackson, L., Motzke, I., Perfecto, I., Whitbread, A., 2012. Global food security, biodiversity conservation, and the future of agricultural intensification. Biol. Conserv. 151 (1), 53-59.

[17] Godfray, C., Beddington, J. R., Crute, I. R., Haddad, L., Lawrence, D., Muir, J. F., Pretty, J., Robinson, S., Thomas, S. M., Toulmin, C., 2010, 'Food security: the challenge of feeding 9 billion people', Science 327, 812-818.

[18] Sachs, J. D., Remans, R., Smukler, S., Winowiecki, L., Andelman, S. J., Cassman, K. G., Castle, D., DeFries, R., Denning, G., Fanzo, J., Jackson, L. E., Leemans, R., Lehmann, J., Milder, J. C., Naeem, S., Nziguheba, G., Palm, C. A., Pingali, P. L., Reganold, J. P., Richter, D. D., Scherr, S. J., Sircely, J., Sullivan, C., Tomich, T. P., Sanchez, P. A., 2010, 'Monitoring the World's agriculture,' Nature 466, 558-560

[19] Sulecki, J., 2018. Association seeks a definitive definition of "precision agriculture"— What's your vote? https://www.precisionag.com/events/association-seeks-definitive-definition-of-precision-agriculturewhats-your-vote/ (accessed 15 November 2018).

[20] Linsley, C.M., and F.C. Bauer. 1929. Test your soil for acidity. Univ. I.L. College Agric. Agric. Exp. Station Circular 346.

[21] Haneklaus, S., and E. Schnug. 2002. An agronomic, ecological and economic assessment of site specific fertilisation. Landbauforsch. Völkenrode 52(3):123-133 https://literatur.thuenen.de/digbib_extern/zi028255.pdf

[22] McBratney A, Whelan B, Thomir A, Bouma J (2005) Future directions of precision agriculture. Precis Agric 6:7-23

[23] Oliver MA, Bishop T, Marchant B (2013) Precision agriculture for sustainability and environmental protection. Routledge, Oxon.

[24] Stafford JV (2000) Implementing precision agriculture in the 21st century. J Agric Eng Res 76(3):267-275

[25] Anderson, D.W., and Gregorich, E.G., 1984. Effect of soil erosion on soil quality and productivity. Soil erosion and degradation. Pages 105-113. Proceedings of 2nd annual western provincial conference on rationalisation of water and soil research and management. Saskatoon, Sask., Canada.

[26] Doran, J.W., Sarrantonio, M., and Lieberg, M.A., 1996. Soil health and sustainability. Adv. Agron. 56: 1-54.

[27] F. Bastida , A. Zsolnay, T. Hernández, C. García Past. 2008. Present and future of soil quality indices: A biological perspective. Geoderma. 147 : 159-171.

[28] Gregorich, L.J. and Acton, D.F., Acton and L.J. Gregorich, eds. 1995. Understanding soil health. The health of our soils-towards sustainable agriculture in Canada. D.F. Pages 5-10. Centre for Land and Biological Resources Research, Research Branch, Agriculture, and Agri-Food Canada, Ottawa, Ont., Canada.

[29] Carter, M.R., Gregorich, E.G., Anderson, D.W., Doran, J.W., Janzen, H.H., Pierce, F.J. 1997. Concepts of soil quality and significance. in: Developments in Soil Science, (Eds.) E.G., Gregorich, M.R. Carter, Vol. Volume 25, Elsevier.

[30] U.S.D.A. 2006. Soil Quality Institute. Natural resources conservation service. Available at

https:// soils.usda.gov/sqi/. (Retrieved on 20 February 2006.)

[31] Ball, A., & De la Rosa, D., In N. Uphoff, A. Ball, E. Fernandes, H. Herren, O. Husson, M. Laing, Ch. Palm, J. Pretty, P. Sanchez, N. Sanginga, & J. Thies (Eds.) Boca Raton, FL: Taylor & Francis/C.R.C. Press. 2006. Modeling possibilities for the assessment of soil systems. Biological approaches to sustainable soil systems: 683-692.

[32] Imeson, A., Arnoldussen, A., De la Rosa, D., Montanarella, L., Dorren, L., Curfs, M., Arnalds, O., & Van Asselen, S. 2006. SCAPE: Soil conservation and protection in Europe. The way ahead. Luxembourg: CEE-JRC.

[33] Nortcliff, S. 2002. Standardisation of soil quality attributes. Agriculture, Ecosystems & Environment. 88: 161-168.

[34] Blum, W.E.H. and Santelises, A.A. 1994. A concept of sustainability and resilience based on soil functions. D.J. Greenland and I. Szabolcs, eds. Soil resilience and sustainable land use: 535-542. C.A.B. International, Wallingford, U.K.

[35] Arshad, M.A., and Coen, G.M., 1992a. Characterisation of soil quality: physical and chemical criteria. Am. J. Alter. Agric. 7: 25-30.

[36] Arshad, M.A., Coen, G.M., 1992b. Characterisation of soil quality: Physical and chemical criteria. American Journal of Alternative Agriculture, 7(Special Issue 1-2), 25-31

[37] Warkentin, B.P., 1995. The changing concept of soil quality. J. Soil Water Cons. 50: 226-228.

[38] Lal, R.; Stewart, B.A. Soil Degradation. 1990. Springer-Verlag: New York, NY, U.S.A.

[39] Myers, N. Gaia. 1993. An Atlas of Planet Management. Anchor/ Double Day: Garden City, NY, U.S.A.

[40] Pimentel, D., C. Harvey, P. Resosudarmo, K. Sinclair, D. Kurz, M. Mcnair, S. Crist, L. Shpritz, L. Fitton, R. Saffouri, R. Blair. 1995a. Environmental and Economic Costs of Soil Erosion and Conservation Benefits SCIENCE .257 (5201): 1117-1123.

[41] Pimentel, D., Harvey, C., Resosudarmo, P., Sinclair, K., Kurz, D., McNair, M., Crist, S., Sphpritz, L., Fitton, L., Saffouri, R. 1995b. Environmental and economic costs of soil erosion and conservation benefits. Science. 267:1117,–1123.

[42] Don, A., Schumacher, J., Freibauer, A. 2011. Impact of tropical land-use change on soil organic carbon stocks – a meta-analysis. Glob. Change Biol. 17:1658,–1670.

[43] Luo, Z., Wang, E., Sun, O.J. 2010. Can no-tillage stimulate carbon sequestration in agricultural soils. A meta-analysis of paired experiments. Agric. Ecosyst. Environ. 139: 224-231.

[44] Hafez, E.E., Elbestawy, E. 2009. Molecular characterisation of soil microorganisms: effect of industrial pollution on distribution and biodiversity. World J Microbiol Biotechnol. 25: 215-224. https://doi.org/10.1007/s11274-008-9881-5

[45] Setälä, H., Bardgett, R.D., Birkhofer, K. 2014. Urban and agricultural soils: conflicts and trade-offs in the optimisation of ecosystem services. Urban Ecosyst 17, 239-253. https://doi.org/10.1007/s11252-013-0311-6

[46] Tsiafouli, M.A., Thébault, E Sgardelis, S.P et al. 2015. Intensive agriculture reduces soil biodiversity across Europe. Glob. Chang. Biol .21: 973-985.

[47] Jie, C., Jing-zhang, C., Man-Zhi, T. 2002. Soil degradation: a global problem endangering sustainable development. *J. Geogr. Sci* .12: 243-252. https://doi.org/10.1007/BF02837480

[48] Andrews, S.S., Carroll, C.R. 2001. Designing a soil quality assessment tool for sustainable agroecosystem management. Ecological Applications, 11(6), 1573-1585.

[49] Chauhan, A., Mittu, B. 2015. Soil Health - An Issue of Concern for Environment and Agriculture. Journal of Bioremediation & Biodegradation, 6, 286.

[50] Chen, S., 1996. The quantitative evaluation of cultivated soil quality in islands of Fuzhou. Journal of Fujian Teachers University (Natural Sciences), 12(3), 84-88.

[51] Doran, J.W., Parkin, T.B., 1994. Defining and Assessing Soil Quality. in: Defining Soil Quality for a Sustainable Environment (Eds.) J.W. Doran, D.C. Coleman, D.F. Bezdicek, B.A. Stewart, S.S.S.A. Madison, WI

[52] Larson, W.E., and Pierce, F.J., 1991. Conservation and enhancement of soil quality. Evaluation for sustainable land management in the developing world. Pages 175-203. I.B.S.R.A.M. Proc, No. 12, vol. 2, Technical papers, Bangkok, Thailand.

[53] Loveland, P.J., Thompson, T.R.E. 2002. Identification and development of a set of national indicators for soil quality. R&D Project Record P5-053/PR/02. Environment Agency, U.K.

[54] Moebius-Clune, B.N., Moebius-Clune, D.J., Gugino, B.K., Idowu, O.J., Schindelbeck, R.R., Ristow, A.J., Van Es, H.M., Thies, J.E., Shayler, H.A., McBride, M.B., Wolfe, D.W., Abawi, G.S. 2016. Comprehensive assessment of soil health. The Cornell framework manual, Edition 3.0. Third Edition. Ed. Cornell University, Geneva, NY.

[55] Velasquez, E., Lavelle, P., Andrade, M., 2007. G.I.S.Q., a multifunctional indicator of soil quality. Soil Biology & Biochemistry, 39(12), 3066-3080.

[56] Wienhold, B.J., Andrews, S.S., Karlen, D.L. 2004. Soil quality: a review of the science and experiences in the U.S.A. Environmental Geochemistry and Health, 26(2-3), 89-95.

[57] Chen, Z.S., 1999. Selecting indicators to evaluate soil quality. Extension Bulletin 473. Food and Fertilizer Technology Center (F.F.T.C.).

[58] Idowu, O.J., van Es, H.M., Abawi, G.S., Wolfe, D.W., Ball, J.I., Gugino, B.K., Moebius, B.N., Schindelbeck, R.R., Bilgili, A.V. 2008. Farmer-oriented assessment of soil quality using field, laboratory, and V.N.I.R. spectroscopy methods. Plant and Soil, 307(1-2), 243-253.

[59] Karlen, D.L., Gardner, J.C., Rosek, M.J. 1998. A soil quality framework for evaluating the impact of C.R.P. Journal of Production Agriculture, 11(1), 56-60.

[60] Kirchmann, H., Andersson, R., 2001. The Swedish system for quality assessment of agricultural soils. Environmental Monitoring and Assessment, 72(2), 129-139

[61] Merrington, G., 2006. The development and use of soil quality indicators for assessing the role of soil in environmental interactions. The Environment Agency.

[62] Schipper, L.A., Sparling, G.P. 2000. Performance of soil condition indicators across taxonomic groups and land uses—Soil Science Society of America Journal, 64(1), 300311.

[63] Seybold, C.A., Herrick, J.E., Brejda, J.J. 1999. Soil resilience: A fundamental component of soil quality. Soil Science, 164(4), 224-234.

[64] Shukla, M.K., Lal, R., Ebinger, M., 2006. Determining soil quality indicators by factor analysis. Soil & Tillage Research, 87(2), 194-204.

[65] Stenberg, B., Pell, M., Torstensson, L. 1998. Integrated evaluation of variation

in biological, chemical, and physical soil properties. Ambio, 27(1), 9-15.

[66] Torstensson, L., Pell, M., Stenberg, B., 1998. Need for a strategy for evaluation of arable soil quality. Ambio, 27(1), 4-8.

[67] Bone, J., Head, M., Barraclough, D., Archer, M., Scheib, C., Flight, D., Voulvoulis, N. 2010. Soil quality assessment under emerging regulatory requirements. Environment International, 36(6), 609-622.

[68] Cotching, W.E., Kidd, D.B., 2010. Soil quality evaluation and the interaction with land use and soil order in Tasmania, Australia. Agriculture Ecosystems & Environment, 137(34), 358-366.

[69] Oberholzer, H.-R., Knuchel, R.F., Weisskopf, P., Gaillard, G. 2012. A novel method for soil quality in life cycle assessment using several soil indicators. Agronomy for Sustainable Development, 32(3), 639-649.

[70] Rutgers, M., Mulder, C., Schouten, A.J. 2008. Soil ecosystem profiling in the Netherlands with ten references for soil biological quality. R.I.V.M. Report 607604009/2008. National Institute for Public Health and the Environment (R.I.V.M.).

[71] Wattel-Koekkoek, E.J.W., van Vliet, M.E., Boumans, L.J.M., Ferreira, J., Spijker, J., van Leeuwen, T.C. 2012. Soil quality in the Netherlands for 2006-2010 and the change compared with 1993-1997: Results of the National Soil Quality Monitoring Network. RIVM Report 680718003.

[72] Wienhold, B.J., Karlen, D.L., Andrews, S.S., Stott, D.E. 2009. Protocol for indicator scoring in the soil management assessment framework (S.M.A.F.). Renewable Agriculture and Food Systems, 24(4), 260-266.

[73] Xu, J.M., Zhang, G.L., Xie, Z.M. 2008. Soil quality index and assessment/consultative system. in: Soil quality of China, (Eds.) Z.H. Cao, J.M. Zhou, Science Press. Beijing.

[74] Armenise, E., Redmile-Gordon, M.A., Stellacci, A.M., Ciccarese, A., Rubino, P. 2013. Developing a soil quality index to compare soil fitness for agricultural use under different managements in the Mediterranean environment. Soil & Tillage Research, 130, 91-98.

[75] Pereira J.D., Baretta. D., Bin. D., et al., 2013. Relationships between microbial activity and soil physical and chemical properties in native and reforested Araucaria Angustifolia forests in the state of São Paulo, Brazil. Revista Brasileira de Ciência do Solo 37(572):586.

[76] Lal, R. 1994a. Methods and guidelines for assessing sustainable use of soil and water resources in the tropics. Soil Management Support Services, Tech Mono 21. Washington (D.C.): Soil Management Support Services. p. 55-63.

[77] Lal, R. 1994b. Methods and guidelines for assessing sustainable use of soil and water resources in the tropics. Soil Management Support Services, Tech Mono 21. Washington (D.C.): Soil Management Support Services. p. 55-63.

[78] Adeoye GO, Agboola AA. 1984. Relationship between soil physical and chemical characteristics and ear-leaf concentration of P, K, Mg, Zn, Fe, Mn, and relative yield of maise in soils derived from sedimentary rocks of southwest Nigeria. Fert Res. 5(1):109-119.

[79] Adeyolanu OD. 2012. Soil quality assessment of some agricultural lands in southwest Nigeria [PhD thesis]. Ibadan (Nigeria): Agronomy Department, University of Ibadan. p. 222.

[80] Harris, R.F., Karlen, D.L., Mulla, D.J. 1996. A Conceptual Framework for

Assessment and Management of Soil Quality and Health in Methods for assessing soil quality. S.S.S.A. Special Publication 49. (Eds.) J.W. Doran, A.J. Jones, Soil Science Society of America

[81] Gregory, P.J Simmonds, L. P. and Pilbeam, C. J. "Soil type, climatic regime, and the response of water use efficiency to crop management," Agronomy Journal, vol. 92, no. 5, pp. 814-820, 2000.

[82] Kay, B. D., & Angers, D. A. (2001). 7 Soil Structure. Soil physics companion, 249.

[83] Mausbach M.J and Seybold, C.A "Assessment of soil quality," in Soil Quality and Agricultural Sustainability, R. Lal, Ed., pp. 33-43, Sleeping Bear Press, Chelsea, Mich, U.S.A., 1998

[84] Yong Liu, Huifeng Wang, Hong Zhang, Karsten Liber, A comprehensive support vector machine-based classification model for soil quality assessment, Soil and Tillage Research, Volume 155, 2016, Pages 19-26, ISSN 0167-1987, https://doi.org/10.1016/j.still.2015.07.006.

[85] Campbell, T.; Dixon, K.W.; Dods, K.; Fearns, P.; Handcock, R. Machine Learning Regression Model for Predicting Honey Harvests. Agriculture 2020, 10, 118.

[86] Saiz-Rubio, V.; Rovira-Más, F. From Smart Farming towards Agriculture 5.0: A Review on Crop Data Management. Agronomy, 2020, 10, 207.

[87] Kaur, S., Pandey, S. & Goel, S. Plants Disease Identification and Classification Through Leaf Images: A Survey. Arch Computat Methods Eng 26, 507-530 (2019). https://doi.org/10.1007/s11831-018-9255-6.

[88] Aguilera, P.A., Fernández, A., Ropero, R.F. et al. Groundwater quality assessment using data clustering based on hybrid Bayesian networks. Stoch Environ Res Risk Assess 27, 435-447 (2013). https://doi.org/10.1007/s00477-012-0676-8

[89] Xiao Huang, Guorui Huang, Chaoqing Yu, Shaoqiang Ni, Le Yu, A multiple crop model ensemble for improving broad-scale yield prediction using Bayesian model averaging, Field Crops Research, Volume 211, 2017, Pages 114-124, ISSN 0378-4290, https://doi.org/10.1016/j.fcr.2017.06.011.

[90] Denis Cornet, Jorge Sierra, Régis Tournebize, Benoît Gabrielle, Fraser I. Lewis, Bayesian network modeling of early growth stages explains yam interplant yield variability and allows for agronomic improvements in West Africa, European Journal of Agronomy Volume 75, 2016, Pages 80-88, ISSN 1161-0301, https://doi.org/10.1016/j.eja.2016.01.009.

[91] Yungang Zhu, Dayou Liu, Guifen Chen, Haiyang Jia, Helong Yu, Mathematical modeling for active and dynamic diagnosis of crop diseases based on Bayesian networks and incremental learning, Mathematical and Computer Modelling, Volume 58, Issues 3-4, 2013, Pages 514-523, ISSN 0895-7177, https://doi.org/10.1016/j.mcm.2011.10.072.

[92] De Rainville, F., Durand, A., Fortin, F. et al. Bayesian classification and unsupervised learning for isolating weeds in row crops. Pattern Anal Applic 17, 401-414 (2014). https://doi.org/10.1007/s10044-012-0307-5

[93] Stanaway, M.A., Mengersen, K.L. & Reeves, R. Hierarchical Bayesian modelling of early detection surveillance for plant pest invasions. Environ Ecol Stat 18, 569-591 (2011). https://doi.org/10.1007/s10651-010-0152-x

[94] Russo T.A., Devineni N., Lall U. (2015) Assessment of Agricultural Water Management in Punjab, India, Using Bayesian Methods. In: Setegn S.,

Donoso M. (eds) Sustainability of Integrated Water Resources Management. Springer, Cham. https://doi.org/10.1007/978-3-319-12194-9_9.

[95] T. Waheed, R.B. Bonnell, S.O. Prasher, E. Paulet, Measuring performance in precision agriculture: CART—A decision tree approach, Agricultural Water Management, Volume 84, Issues 1-2, 2006, Pages 173-185, ISSN 0378-3774, https://doi.org/10.1016/j.agwat.2005.12.003.

[96] Xueli Liu, Bingchan Li, Dongqin Shen, Jie Cao, Bo Mao, Analysis of Grain Storage Loss Based on Decision Tree Algorithm, Procedia Computer Science, Volume 122, 2017, Pages 130-137, ISSN 1877-0509, https://doi.org/10.1016/j.procs.2017.11.351.

[97] R.H. Bosma, Le Thanh Phong, Uzay Kaymak, J. Van Den Berg, H.M.J. Udo, M.E.F. Van Mensvoort, Le Quang Tri, Assessing and modelling farmers' decision-making on integrating aquaculture into agriculture in the Mekong Delta, N.J.A.S. - Wageningen Journal of Life Sciences, Volume 53, Issues 3-4, 2006, Pages 281-300, ISSN 1573-5214, https://doi.org/10.1016/S1573-5214(06)80010-2.

[98] Moonjun R., Farshad A., Shrestha D., Vaiphasa C. (2010) Artificial Neural Network and Decision Tree in Predictive Soil Mapping of Hoi Num Rin Sub-Watershed, Thailand. In: Boettinger J.L., Howell D.W., Moore A.C., Hartemink A.E., Kienast-Brown S. (eds) Digital Soil Mapping. Progress in Soil Science, vol 2. Springer, Dordrecht. https://doi.org/10.1007/978-90-481-8863-5_13

[99] Kim, K., Yoo, K., Ki, D. et al. Decision-Tree-based data mining and rule induction for predicting and mapping soil bacterial diversity. Environ Monit Assess 178, 595-610 (2011). https://doi.org/10.1007/s10661-010-1763-2

[100] Rossi Neto, J., de Souza, Z.M., de Medeiros Oliveira, S.R. et al. Use of the Decision Tree Technique to Estimate Sugarcane Productivity Under Edaphoclimatic Conditions. Sugar Tech 19, 662-668 (2017). https://doi.org/10.1007/s12355-017-0509-7

[101] Priyanka T, Soni P, Malathy C. Agricultural Crop Yield Prediction Using Artificial Intelligence and Satellite Imagery. Eurasian Journal of Analytical Chemistry. 2018;13(S.P.): emEJAC181194.

[102] Daniel J., Andrés PU., Héctor S., Miguel B., Patrick V.D., Marco T. (2008) A Survey of Artificial Neural Network-Based Modeling in Agroecology. In: Prasad B. (eds) Soft Computing Applications in Industry. Studies in Fuzziness and Soft Computing, vol 226. Springer, Berlin, Heidelberg

[103] Kirtan Jha, Aalap Doshi, Poojan Patel, Manan Shah, A comprehensive review on automation in agriculture using artificial intelligence, Artificial Intelligence in Agriculture, Volume 2, 2019, Pages 1-12, ISSN 2589-7217, https://doi.org/10.1016/j.aiia.2019.05.004.

[104] Monisha Kaul, Robert L. Hill, Charles Walthall, Artificial neural networks for corn and soybean yield prediction, Agricultural Systems, Volume 85, Issue 1, 2005, Pages 1-18, ISSN 0308-521X, https://doi.org/10.1016/j.agsy.2004.07.009.

[105] Kennedy Were, Dieu Tien Bui, Øystein B. Dick, Bal Ram Singh, A comparative assessment of support vector regression, artificial neural networks, and random forests for predicting and mapping soil organic carbon stocks across an Afromontane landscape, Ecological Indicators, Volume 52, 2015, Pages 394-403, ISSN 1470-160X, https://doi.org/10.1016/j.ecolind.2014.12.028.

[106] Behnaz Nahvi, Jafar Habibi, Kasra Mohammadi, Shahaboddin Shamshirband, Othman Saleh Al Razgan, Using self-adaptive evolutionary algorithm to improve the performance of an extreme learning machine for estimating soil temperature, Computers and Electronics in Agriculture, Volume 124, 2016, Pages 150-160, ISSN 0168-1699, https://doi.org/10.1016/j.compag.2016.03.025.

[107] Shinji Fukuda, Wolfram Spreer, Eriko Yasunaga, Kozue Yuge, Vicha Sardsud, Joachim Müller, Random Forests modelling for the estimation of mango (*Mangifera indica* L. cv. Chok Anan) fruit yields under different irrigation regimes, Agricultural Water Management, Volume 116, 2013, Pages 142-150, ISSN 0378-3774, https://doi.org/10.1016/j.agwat.2012.07.003.

[108] Aurore Philibert, Chantal Loyce, David Makowski, Prediction of N2O emission from local information with Random Forest, Environmental Pollution, Volume 177, 2013, Pages 156-163, ISSN 0269-7491, https://doi.org/10.1016/j.envpol.2013.02.019.

[109] Jinyoung Rhee, Jungho Im, Meteorological drought forecasting for ungauged areas based on machine learning: Using long-range climate forecast and remote sensing data, Agricultural and Forest Meteorology, Volumes 237-238, 2017, Pages 105-122, ISSN 0168-1923, https://doi.org/10.1016/j.agrformet.2017.02.011.

[110] Inacio H. Yano, Jose R. Alves, Wesley E. Santiago, Barbara J.T. Mederos, Identification of weeds in sugarcane fields through images taken by UAV and Random Forest classifier, IFAC-papers online, Volume 49, Issue 16, 2016, Pages 415-420, ISSN 2405-8963, https://doi.org/10.1016/j.ifacol.2016.10.076.

[111] Saussure, S., Plantegenest, M., Thibord, J. et al. Management of wireworm damage in maise fields using new, landscape-scale strategies. Agron. Sustain. Dev. 35, 793-802 (2015). https://doi.org/10.1007/s13593-014-0279-5.

[112] Everingham, Y.L., Lowe, K.H., Donald, D.A. et al. Advanced satellite imagery to classify sugarcane crop characteristics. Agron. Sustain. Dev. 27, 111-117 (2007). https://doi.org/10.1051/agro:2006034.

[113] Debasmita Misra, Thomas Oommen, Avinash Agarwal, Surendra K. Mishra, Anita M. Thompson, Application and analysis of support vector machine-based simulation for run-off and sediment yield, Biosystems Engineering, Volume 103, Issue 4, 2009, Pages 527-535, ISSN 1537-5110, https://doi.org/10.1016/j.biosystemseng. 2009.04.017.

[114] Miloš Kovačević, Branislav Bajat, Boško Gajić, Soil type classification and estimation of soil properties using support vector machines, Geoderma, Volume 154, Issues 3-4, 2010, Pages 340-347, ISSN 0016-7061, https://doi.org/10.1016/j.geoderma.2009.11.005.

[115] Tisen Huang, Rui Yang, Wenshan Huang, Yiqi Huang, Xi Qiao, Detecting sugarcane borer diseases using support vector machine, Information Processing in Agriculture, Volume 5, Issue 1, 2018, Pages 74-82, ISSN 2214-3173, https://doi.org/10.1016/j.inpa.2017.11.001.

[116] Koji Kawamura, Laurence Hibrand-Saint Oyant, Fabrice Foucher, Tatiana Thouroude, Sébastien Loustau, Kernel methods for phenotyping complex plant architecture, Journal of Theoretical Biology, Volume 342, 2014, Pages 83-92, ISSN 0022-5193, https://doi.org/10.1016/j.jtbi.2013.10.016.

Chapter 10

# Artificial Intelligence and Water Cycle Management

*Carmine Massarelli, Claudia Campanale and Vito Felice Uricchio*

## Abstract

Artificial intelligence applications play a crucial role in improving environmental quality from all points of view. Digital technologies have revolutionized our way of life as they are permeated to a capillary level in our daily life. On the other hand, the data produced every second cannot be managed by a human mind due to a certain physical and temporal impossibility, so artificial intelligence, algorithms written by men to perform human reasoning, they can accomplish this arduous task. In this chapter we will address the potential of artificial intelligence to process important amounts of data and analyze existing relationships also through a focus on the conservation capacity of one of the most precious resources: water.

**Keywords:** water cycle, artificial intelligence, smart communities, collective knowledge, big data

## 1. Introduction

When we talk about Artificial Intelligence, we immediately think of cutting-edge technologies, robots capable of understanding and deciding the actions to be taken and a futuristic world in which machines and humans coexist. In fact, Artificial Intelligence and its use are much more real than one might imagine and are now used in various areas of daily life. In this chapter, we want to discuss the state of the art of the application of artificial intelligence for the management of water resources.

The "White Paper on Artificial Intelligence - A European Approach to Excellence and Trust", COM (2020) 65, highlights how digital technology has improved our lives allowing easier access to knowledge and content. Nowadays, Europe is called to make two transformations (green and digital) which, in the water management sector, could have common objectives: the first requires to take actions towards more sustainable solutions, the second consists in directing the social transformations in such a way that every citizen can take full and maximum benefit. In line with these objectives, the applications of artificial intelligence can contribute to preserving the environment and first of all the most precious resource: water.

Digital and AI are an engine of change that can allow companies to expand and consolidate their competitive positions in international markets in the name of sustainability [1]. The challenges faced in recent months and the Commission's guidelines [2] aim to promote a series of initiatives, both legislative and development programs, to guide our society towards a more modern, equitable model that can exploit better the power of data and AI. One focus will be on extracting hidden information from available data.

The full exploitation of the significant potential of AI in the water sector to process important amounts of data and analyze relationships allows supporting technical and political choices especially during the planning stage. The management of water resources is particularly important also for the protection of the natural biodiversity which expresses a profound complexity, which is reflected in the extraordinary numbers of animal and plant biodiversity and the environmental parameters that our territory records [3]; but also natural and anthropogenic threats touch different levels of scale and complexity, causing alterations and changes in the stability of ecosystems, reducing functionality and resilience. Mathematical and geostatistical tools for the study of environmental complexity represent a fundamental tool for understanding the complexity of processes that can impact on the quality of life, but many times they are not sufficient [4]. Understanding and analyzing the complex and often imperceptible relationships between the environment and health are fully part of the issues that require a joint scientific commitment which, starting from the in-depth examination of each environmental and anthropic component in its complexity numerical, leads to an indispensable multidisciplinary collaboration often difficult to achieve [5]. A multitude of algorithms defined by different expertise and formalized in Artificial Intelligence systems can contribute to overcoming these criticalities [6]. We collectively need integrated analysis strategies of environmental information, which take us beyond the short-term horizon of specific sectoral knowledge, albeit specialized, aiming at the harmony of knowledge to face the challenges of protecting the water resources that impact not positively on the conservation of the environment and the preservation of health. The numbers involved allow us to understand the indispensability of *Big Data* analysis with AI systems: from the 1.000 billion bacterial species that populate the planet and/or our body to the 100.000 chemical compounds that we disperse into the environment and that they reach our organs, to the complexities that each of these elements brings with it individually and in their interaction.

The aforementioned complexity makes the creation of an ecosystem of excellence based on AI extremely positive, capable of introducing scientific innovation that can be transferred to the sector of Public Administration and Companies.

The hope is that the economic support for research can stimulate and reward the excellences present in the territories, determining a distribution growth of the communities on AI. I believe it is appropriate to encourage the use of technologies based on AI in relation to the resolution of application problems in sectors based on regional strategies designed for innovation areas such as "Human health and the environment", as well as "Sustainable manufacturing" with particular attention reference to water management in production processes.

In particular, concerning the theme of "*Human and environmental health*", Artificial Intelligence can make a decisive contribution in terms of territorial control through automatic image classification systems (*CBIR - Content-Based Image Retrieval System*) that allow using mathematical models, computer implementations of the content of an image, to simulate the principles of the human visual system and to interpret the scenes with a semantics capable of recognizing predefined situations. Such AI applications allow to protect the privacy (as the videos are, in the first instance, analyzed by machines) and to recognize illegal acts such as *spills of wastewater, disposal of waste solids or liquids, picking activities unfair contract and infringements environmental of any kind* that can hurt the environment and in particular on water resources. Similar paths can be used with AI approaches through the application of *semi-automatic Change detection algorithms* functional to the evaluation of territorial transformations, enhancing the significant availability of satellite images acquired by the numerous sensors onboard satellite platforms. In this sense, the environmental and territorial applications affecting the water sector refer to the following areas: illicit disposal, illegal building, land-use change, forest fragmentation, urban growth, loss of agricultural land, availability of resources water supply in lakes and reservoirs, melting of glaciers, the evolution of watercourses, etc.

## 2. Materials and methods

### 2.1 Artificial Intelligence and integrated management of the water cycle

Water, as a primary source of life and as a natural, cultural, economic and political resource, requires *intelligent management* and the same artificial intelligence can assist in the involvement of the collective intelligence dispersed in citizenship, now evolved into a true *Smart Communities*, to ensure the protection, conservation and rational and optimal use in an *Adaptive Water Management* regime.

The management of water resources requires the formulation of new paradigms capable of combining, on the one hand, the protection of water resources, through new systems and intelligent technologies, capable of increasing the efficiency in the use of resources and the performances of networks and treatment plants present in the territory and on the other hand the development of new monitoring systems distributed and easy to access for widespread control of the quality status. In both cases, the AI plays an extraordinarily important role, especially in the presence of massive amounts of data: an increasingly recurring situation due to the strengthening of water and environmental monitoring systems. The development of interoperable technologies capable of promoting the dissemination and exchange of large volumes of information between decision-makers, managers and citizens, can lead to the creation of *widespread knowledge* capable of feeding artificial intelligence systems and aimed at supporting better environmental protection, ending with a direct impact on the educational and behavioral side. In this direction, the *ubiquity of water*, in every declination of social and productive life, constitutes a natural element to channel information and to consolidate a new culture that can combine the expressions of AI favoring growth, the sharing of structured expert knowledge and not and increasing the sense of belonging to one's territory and to the natural resources it expresses. The recognition of water, as a human right, passes through the acceptance of the sense of widespread (public) ownership and responsibility that must guide both the small daily choices and the big planning, management, political and administrative decisions.

In line with the definitions of the Water Framework Directive 2000/60/EC and the updates in progress, and in general with the articulated Community, national and regional regulatory framework, it is necessary to pursue the objectives of safeguarding, protecting and improving the environmental quality of water bodies, as well as the prudent and rational use of natural resources based on management that is not only sustainable but adaptable to the circumstances that arise also as a result of global changes: all elements of high complexity that find in AI an indispensable ally. In this vision, participatory processes that can also be activated through AI are crucial for triggering paths that lead to the construction of the economic and social vocation of smart cities.

The keyword underlying the concept of AI is "*integration*" to be achieved at various levels of knowledge: both in the management of the entire " *water supply chain* " but also with the active involvement of citizens, management bodies, research bodies and universities, companies, supervisory authorities, to achieve management of water resources capable of facing complexities, in line with the needs of environmental sustainability and reduction of impacts.

A correct understanding of the management of water resources can certainly not be limited to the simple government of only one of the components such as, procurement, distribution networks, purification, etc., but it requires a broader perspective that allows the analysis and definition of coordinated and integrated strategies that affect the entire water cycle (**Figure 1**).

Furthermore, even the potentially most efficient strategies have no chance of success if they are not supported by an "awareness" of citizens who must be directly involved as actors within a system that cannot ignore virtuous behavior at the macro level and micro-communities.

It must be emphasized that the constantly growing demographic evolutions, the consequent increased use of the intensification of crops, the effects of climate change with the increase in the frequency of extreme events, determine an extreme urgency in implementing every possible solution (including technological and "intelligent") that can make the resource management system as a whole more efficient, both in quantitative and qualitative terms.

In this context, the research activities on AI that envisage water management in the implementation of environmental policies, in close connection with the Europe 2020 strategy which has identified smart growth, sustainable growth and inclusive growth, as engines of the relaunch of the economy.

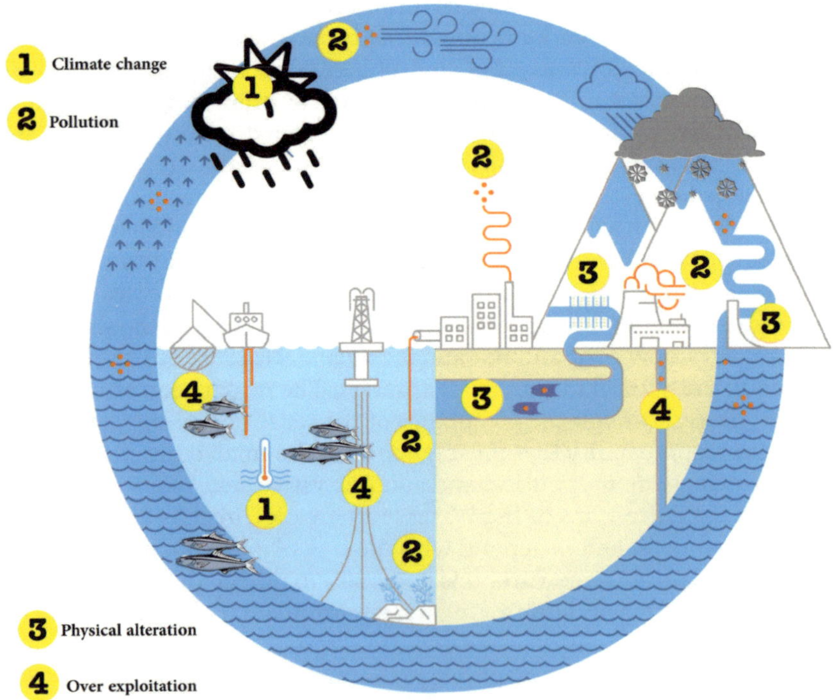


**Figure 1.**
*Water cycle and pressure factors and areas of application of AI [7]. Legend: 1) climatic change, 2) pollution, 3) physical alterations, 4) over exploitation.*

The applications of AI in the water management sector, operating on huge amounts of data, concern the monitoring and management of extreme events also interface with the "*Territory security*" area, while others concern the collection and storage of data, their dissemination and their interoperable using interfaces with the "*Home automation and Smart Grids*" area, in particular concerning aspects relating to the improvement of the quality of life in domestic environments, the reduction of management costs and the transmission of information through *Power Line Communication* (PLC) and their storage using Cloud technology [8].

By way of example, a monitoring system based on AI technologies makes it possible to more effectively target control actions on diffused loads generated as a result of an overflow of the network and on production ones, in order to reduce the presence of metal contaminants and organic and maximize nutrient recovery. The AI itself, combined with innovative devices for controlling the efficiency of urban sewers, allows immediate intervention, reducing the risk of contamination of the unsaturated and groundwater [9]. Precisely for these reasons, these technologies are particularly functional for achieving *good ecological and chemical status* in water bodies, envisaged by the European Directive 2000/60/EC. In response to what is strongly desired by administrators, managers and citizens, the AI itself uses *early warning indicators* that make it possible to identify and suggest mitigation strategies on a local scale of extreme events attributable to natural factors (e.g. climate change and consequent changes in the regime rainfall) or anthropogenic (eg illicit disposal or accidental spills).

AI can be decisive in identifying and managing adaptation guidelines in relation to the climate changes underway. In particular, it can be useful for:

- water supply management sector:
  - **control of leaks,** orienting measurement strategies and priorities and the most effective types of intervention to reduce water dispersion;
  - the definition of **investments in water networks and infrastructures**, supporting a holistic water policy that takes into account an extremely large number of technical, managerial, social and economic variables;
- water resources management sector:
  - the promotion of the **natural conservation of water** by orienting the areas in which to favor it both for employment opportunities and for the reduction of hydraulic risk;
  - **the aggregation** of fragmented **surveillance activities** between the different management and control bodies, also in order to improve the quality and use of information;
  - support **capacities in adapting to extreme climatic events**, in particular as regards the control of floods and drought;
- transversal sectors for example in relation to climate change:
  - **the efficiency of water use in all sectors** and guaranteeing sustainable withdrawal and supply of freshwater also to reduce conflicts of use and to address water scarcity in the short, medium and long term;

- support and promote **inter-sectoral, regional, national and sub-national policies** on water management and quality to increase the resilience of water supply, treatment, storage and transport systems as well as hygiene systems, ensuring adequate knowledge and implementation of hygiene practices;

- support the adoption and implementation of a **risk-based approach in the water** and sanitation sector (ie water safety plans, sanitation safety plans), including management of data on diseases, the design of early warning systems based on projections of distributions pathogens, emerging chemical contaminants and/or subject to ordinary control;

- support the modeling and monitoring of hazards, including algae blooms and the production of toxins in the aquatic environment;

- avoid the effects on water quality due to floods, etc.

As better specified below, in urban and semi-urban areas, AI can also intervene effectively in the urban wastewater purification sector, orienting technological applications to improve the efficiency and versatility of plants and favoring low environmental impact technologies, in terms of occupied surfaces, production of sludge and odor emissions, aimed at maximizing energy recovery and the recovery of raw materials and in particular nutrients and biofuels [10].

## 2.2 Smart devices in the home: data source for AI applications

The availability of sensors in the home, interfaced via Wi-Fi network to routers or smartphones and computers, allows to acquire important amounts of data that can be used both for the benefit of the individual user, but also and above all for the benefit of the manager, allowing responsible management of consumption, maintenance of the plants and networks as well as the operating pressures in the different time bands.

In addition, the *intelligent* management of water distribution systems allows ample space for the introduction of innovations in the name of water-saving and environmental sustainability, obtaining useful advantages in terms of monitoring and optimization of resources.

Technologies based on microelectronic applications make it possible to create multiple systems of specialized micro and nanosensors, capable of monitoring in real-time the main physic-chemical parameters that establish the characteristics of the water.

Among the many types of sensors available or in advanced development we remember in particular:

- selective ion field-effect transistors (ISFET) and enzyme-modified field-effect transistors (ENFET) for the measurement of pH, concentration of nitrates and ions of alkali metals and halides such as Ca and Cl, of surfactants anionic and cationic, pesticides and for monitoring the level of fertilizers in the soil;

- potentiometric sensors with a polymeric membrane with selective ionic planar electrodes used for the determination of the presence of organic ionic pollutants;

- potentiometric sensors based on amorphous chalcogenides for the detection of the presence of heavy metals, including Cu, Pb, Cd, Ag, Cr and Fe, even at very low concentrations of the order of nanomoles;

- semiconductor/graphene/metal (SGM) thin-film devices for the detection of traces of organic contaminants;

- laser interferometry sensors that measure the change in the refractive index of water with respect to the reference value determined by the presence of chemical contaminants, capable of detecting the main chemical contamination agents at the level of one part per million and acting as localized *early warning* systems;

- MEMS (Micro- Electro - Mechanical Systems) acceleration sensors consisting of micrometric mechanical transducers to measure the variations in water flow with an integrated wireless data transmission system with very low consumption;

- arrays of MEMS sensors (Micro- Electro - Mechanical Systems) with enzymatic amplification for the detection of bacterial agents by means of amperometric techniques.

The dimensions of these sensors, all of the order of more than a few millimeters, allow them to be installed in smart meters and, in perspective, directly in the flow limiters of the taps according to advantages (water conservation, energy saving) and disadvantages (initial investments, reflective surfaces and extremely bright colors for infrared sensors) [11] also from the point of view of the loss of transmission signals [12].

For data collection, industrial research has already developed numerous types of computational models [13], which, however, are susceptible to important innovations related to the measurement of consumption with quantitative assessments. Sensors installed in the same meters have to transmit data wirelessly to second-level data collection systems, similar to the cells of the cellular telephone network, in turn, connected directly to the data collection and processing network of the water network operator. Similarly, techniques borrowed from artificial intelligence are mature that can allow data to be collected from smart meters to transmit them (after appropriate processing) directly to the network manager, transparently using the decentralized network made up of the smartphones of users who are nearby. This could avoid the implementation of second-level data collection systems, with significant benefits at the level of complexity and overall cost of the system. Such projects, but on a much smaller scale, have already been developed and implemented both in the Netherlands and in Singapore. A Dutch water distributor, has implemented a "*smart grid*" of sensors for real-time monitoring of water quality at a chemical and bacteriological level, considering, in particular, the presence of pesticides, hormones and pharmaceutical products. The sensors are developed are mounted in flow cells crossed by the distribution water. But research continues with the development of increasingly innovative technological solutions and increasingly providing useful information to artificial intelligence systems [14].

The chemical-bacteriological characteristics are monitored in real-time by measuring the change in the refractive index of the water using a laser beam and comparing it with the seasonal reference values for pure water. All this makes it possible to build an "*early warning*" system capable of dealing with water contamination events in real-time within the macroscopic grid made up of the installed sensors. The system incorporates wireless transmission modules that allow to automatically transmit the measured data to the operator. In Singapore, the local water manager, the Public Utility Board (PUB), has also implemented a similar *smart grid* using the same sensors [14].

Ultimately, these low-cost *smart devices* make it possible [12] to achieve significant water savings through the active involvement of citizens. These devices, suitably miniaturized and customized, allow total integration with the innovative information transmission systems as well as capturing the energy necessary for self-supply.

These low-cost sensors are born with the aim of connecting element between the *Smart Communities* and the *intelligent government of resources*, allowing each citizen to be an active part in the acquisition of distributed information can be used both directly (through specific *apps*) and both with the mediation of interoperable Artificial Intelligence systems that make it possible to return the information with high added value to managers and citizens.

The implementation of low-cost sensors that can be marketed through distribution channels of simple access, allows reaching citizens in a widespread manner, supporting mechanisms for acquiring information useful both to users/ citizens and to public decision-makers. In this direction, AI systems capable of reading and interpreting large amounts of data from individual users allows *for precision water management* that can be commensurate with each individual user or even aggregated by the district.

The AI System, acquiring information from individual users on a daily basis, is able to identify any anomalies in real-time, signaling possible malfunctions and providing alarm signals. In this direction, AI can be profitably aimed at implementing efficient systems for controlling and reducing water losses both in distribution networks and in the home, favoring technological convergences between the scientific fields of electronics and hydraulics.

## 2.3 AI in the monitoring of urban waste

The monitoring of discharges assumes particular importance in the context of water management, both because it is allowed to obtain useful information along the pipeline in order to evaluate the presence of any illegal connections and illegal discharges and both because it allows modulating the management of the plant's purification as a function of the monitored polluting load and other ancillary parameters. The monitoring of the hydraulic efficiency of the drainage system,

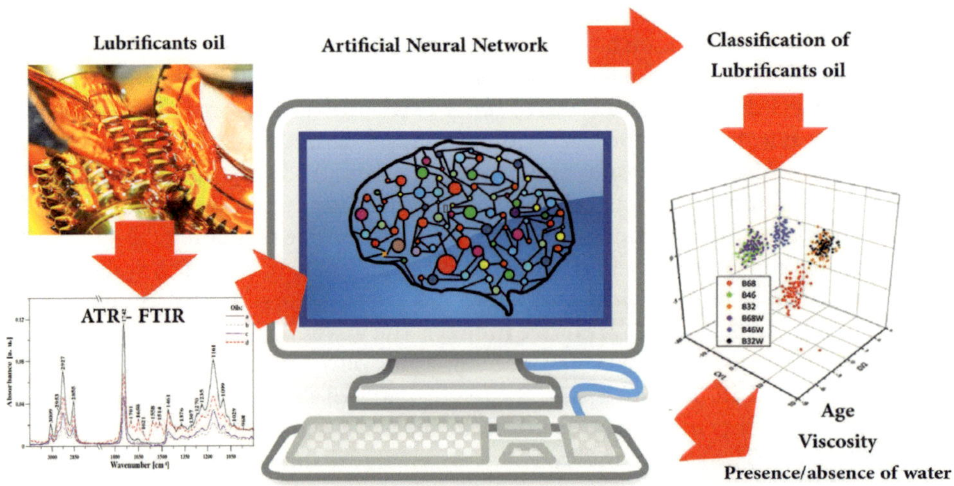


**Figure 2.**
*Artificial intelligence applications for the expeditious characterization of oils in water [15].*

the chemical and physical parameters and the functionality of wastewater treatment plants is an important prerequisite for ensuring the smooth operation on environmental, health, a city economic and social.

In this field too, the main applications of AI derive from the presence of sensors and the progress of research. Engineers and chemists have made it possible to develop devices capable of evaluating pollutants present in water such as oils, hydrocarbons and/or derivatives from a qualitative and quantitative point of view (**Figure 2**) [15].

In particular, starting from the acquisition of detailed information on the quality and quantity of wastewater that passes through a sewer section, it is possible to obtain useful data referring to events that characterize the functioning of a wastewater collection system (variations in flow rates and load, anomalous discharges, exceptional meteorological events). On the basis of the data collected from chemical/physical monitoring, with the help of AI, it is possible to develop useful knowledge to build a complete picture on the composition of wastewater, also identifying the presence of inter-correlation between the different parameters and users (private, artisanal or industrial) that contribute to the composition of the wastewater. Elements of innovations and which contribute to confer added value to Artificial Intelligence applications mainly relate to:

- optimization processes of qualitative-quantitative monitoring of the wastewater collected in the sewer pipes;

- determination of primary parameters (directly measured by existing sensors) and secondary (recognizable through specific patterns) and the definition of their weight on the analysis of the wastewater collected in the sewer pipes;

- progressive refinement of typical artificial intelligence methodologies and numerical resolution of complex and approximate systems of equations to study the correlation between the data acquired by sensors and the data derived from chemical/physical analytical monitoring;

- definition of a model the applicability on a sliding scale to the continuous monitoring and in real-time of the wastewater in the sewage pipelines collected;

- system realization of the alert and who cannot afford to intervene rapidly and in a targeted manner by providing timely, useful information both on-site and on the modalities and intervention issues.

The monitoring of sewage discharges and the intelligent management of data with the consequent construction of scenarios, also allows the optimization of biological purification processes also for the purpose of subsequent reuse of the effluent of the purification plants. For example in the agricultural sector, within a broader framework of guaranteeing food safety (ensuring quality agricultural production), reducing hydrological stress in the summer (characterized by scarcity of irrigation water of natural origin), reducing pollution of surface and groundwater (reducing the excess of nutritional elements that flow into the surface water network and decreasing the pollution of the groundwater by nitrates).

The possibility that non-authorized industrial and/or artisanal discharges occur in the sewerage network, with high concentrations of chemical substances, represents a criticality that often occurs and that can affect subsequent purification processes but also of circular economy (both with reference to waters than mud).

In relation to the type, to the masses and concentrations of quests and chemicals, in fact, such recovery processes may be more or less efficient or even be inhibited. The identification and subsequent elimination of unauthorized discharges are therefore essential for the success of nutrient recovery processes and can be carried out through the combination of modeling and qualitative-quantitative monitoring of the sewer network. It should be noted that these discharges have the characteristic of being intermittent and irregular over time and can also occur in points other than those of the production activities that generated them. Their identification is therefore complex and is difficult to detect through ordinary sampling and analysis methods. However, the availability of smart sensors interfaced with AI systems can gradually refine their localization and therefore selectively organize and improve the control activity (also by modifying the location of the sensors) until the exact identification (even in flagrant) of the unloading operations.

Further monitoring element concerns the sediments into the sewer, which is a very important problem because of the considerable hydraulic and environmental uncertainties associated with the deposits. The accumulation of sediments in the sewer can, in fact, cause considerable hydraulic problems connected to the reduction of the flow capacity of the canals and, consequently, to the increase in the risk of flooding in urban areas; it can also be the cause of significant environmental and health problems, due for example to the resuspension from the bottom of the channels of solids and associated pollutants with consequent discharge through the overflow devices during the most intense meteoric events. In addition, phenomena of anaerobic transformation may occur, linked to the establishment of septic conditions within the accumulations of solid material, with the development of corrosive phenomena, but also with the formation and release of toxic substances and bad smells. Furthermore, the development of management methods of sewage sediments that guarantee a regular solid flow to the treatment allows to optimize the management of purification processes and, at the same time, to act on some of the criticalities that are typically induced by the provision of rainwater on the functionality of urban purification plants. Even with reference to these critical issues, AI support can be strategic.

## 3. Results and discussion

### 3.1 The perspectives of AI in the water sector

The AI accelerate the design of systems procurement, distribution, treatment and reuse of water, using an increasingly widespread use of computer technology and equipment monitoring. Advanced diagnostic tools make water management more customized and intelligent in the water sector. In addition the, in fact, a raised will allow you to overlay information and animations on real-world images with model projections arising from AI applications to help in activities on an daily management and planning and to manage the resource more efficiently. The *Virtual reality* (VR) can make "viewable" projections and modeling predictions on the trend basis and patterns of use of water resources or on climatic scenarios assumed, the *augmented reality* (AR), however, superimposes information generated by a computer to the real world, in quick time. The AI facilitates the integration of these worlds, analyses the incoming data stream, managing large information relating to the scene and superimposes to do it with big data, images or animations relevant, also in 3D. In the near future, we will be able to visualize the system we are imagining to design with the possibility of visualizing the efficiency of use of the different scenarios. Engineers, chemists, biologists, *designers,* etc. they will have new tools to develop collaborative projects, involving expert technicians and young professionals and evaluating the results of design choices in different scenarios of use.

In the coming years, the offer of AI directed to researchers and companies will be able to expand further thanks to programs that are simple to use, to be used in the design, promoting the so-called "fourth industrial revolution": a systemic transformation that can have direct impacts also in the management of the waters.

The AI can determine the output of the information correct exactly at the moment when it is needed, such as when it is necessary to make choices, reducing the chances of error and, increases efficiency and improves the productivity.

Contextualizing to the water sector, the intervention of the AI will be able to optimize the distribution or disposal of water, make the removal of polluted substances more efficient [16], facilitate the reuse of purified wastewater [17], providing real-time images of the areas in which criticalities occur.

In addition, in the last years, artificial intelligence has started to increase the efficiency of the design and synthesis phases of new materials that can also be used in the water sector, making applications faster, easier and more economical, for example, by reducing the use of chemicals or sludge.

In AI, evolutionary machine learning algorithms analyze all relevant experiments; both those that worked and those that failed, effectively preventing further possibilities for error. On the basis of the experiments carried out and the consequent success, the algorithms foresee potentially useful paths. There is no machine learning tool capable of doing all this alone, but AI-based technologies are also spreading in the design of systems and structures for water management such as reservoirs, adductors, lifting systems, distribution, sewer networks, purification plants, etc.

The management of the integrated water cycle is transversal to numerous scientific fields and artificial intelligence is also expanding in all life sciences because it helps to identify *patterns* in complex data sets [18]. The water sector, in particular, allows for huge amounts of data with which to train algorithms, offering significant development opportunities. In fact, artificial intelligence is very successful when there is the possibility of a *training set* of particularly relevant dimensions. The *deep learning* and artificial intelligence tools are amazingly powerful that will provide important answers, especially when interfaced to *smart Technologies* able to acquire data in multiple areas of the integrated water cycle [19].

Even more promising prospects derive from quantum computer applications, which in a few years could greatly exceed the performance of the classical ones, thanks to the significant work on specific hardware and algorithms, exploiting quantum mechanics to perform the calculations and returning greater and further the force on AI.

## 4. Conclusions

The AI may soon create a new form of superintelligence life. Nevertheless also in the field of water management is useful to plan a close relationship synergy between human and artificial intelligence so that it becomes a useful ally, using a shared ethical approach, transparency of governance of innovation processes and the possibility to citizens to exercise their rights and express their opinions by contributing to the growth of artificial consciousness and collective knowledge.

## Author details

Carmine Massarelli*, Claudia Campanale and Vito Felice Uricchio
Water Research Institute,National Research Council of Italy, Bari, Italy

*Address all correspondence to: carmine.massarelli@ba.irsa.cnr.it

## References

[1] Sachs J. D., Schmidt-Traub G., Mazzucato M., Messner D., Nakicenovic N., Rockström J. (2019), Six Transformations to achieve the Sustainable Development Goals. Nature Sustainability, 8.

[2] COM(2020) 65 WHITE PAPER On Artificial Intelligence - A European approach to excellence and trust, Bruxelles, 2018.

[3] Bratsberg, B., & Rogeberg, O. (2018). Flynn effect and its reversal are both environmentally caused. Proceedings of the National Academy of Sciences of the United States of America, 115(26), 6674– 6678. https://doi.org/10.1073/pnas.1718793115

[4] Allen, N. J., & Barres, B. A. (2009, February 5). Neuroscience: Gliamore than just brain glue. Nature. Nature. https://doi.org/10.1038/457675

[5] Campos de Mendonça C.M., Valente de Andrade A.M. (2018), Dynamic Capabilities and Their Relations with Elements of Digital Transformation in Portugal. Journal of Information Systems Engineering & Management, 3.

[6] Athey, S. (2017), "The impact of machine learning on economics" in Ajay K. Agrawal, Joshua Gans, e Avi Goldfarb (eds), The Economics of Artificial Intelligence: An Agenda Economics of Artifi- cial Intelligence, Chicago, IL: University of Chicago Press.

[7] EEA Signals 2018 – Water is life. ISSN: 2443-7662, DOI: 10.2800/52469

[8] Blut M., Wang C., Schoefer K. 2016. Factors influencing the acceptance of self-service technologies. Journal of Service Research, 19, 4, 396-416.

[9] Bogataj D., Bogataj M., Hudoklin D. (2017), Mitigating risks of perishingable products in the cyber-physical systems based on the extended MRP model. International Journal of Production Economics, 193, p. 51-62.

[10] Huang, M., Ma, Y., Wan, J., Chen, X., 2015. A sensor-software based on a genetic algorithm-based neural fuzzy system for modelling and simulating a wastewater treatment process. Appl. Soft Comput. J. 27, 1-10.

[11] Valentini, F. (2019). Smart Electrochemical Portable Tools for Cultural Heritage Analysis: A Review. Sensors 2019, 19, 4303; doi:10.3390/s19194303

[12] Carminati M, Turolla A., Mezzera L., Di Mauro M., Tizzoni M., Pani G., Zanetto F., Foschi J., Antonelli M. 2020. A Self-Powered Wireless Water Quality Sensing Network Enabling Smart Monitoring of Biological and Chemical Stability in Supply SystemsSensors 2020, 20, 1125; doi:10.3390/s20041125

[13] Ashby, F. G., & Waldschmidt, J. G. (2008). Fitting computational models to fMRI. Behav.Res.Methods, 40(1554-351X (Print)), 713– 721.

[14] https://www.dutchwatersector.com/news/optiqua-deploys-80-eventlab-water-quality-sensors-for-vitens-smart-supply-network-the. Last accessed 2021-03-04

[15] Chimeno-Trinchet, C; Murru, C; Diaz-Garcia, ME ; Fernandez-Gonzalez, A ; Badia-Laino, R - Artificial Intelligence and Fourier-transform infrared spectroscopy for evaluating water-mediated degradation of lubricant oils, 2020. DOI: 10.1016/j.talanta.2020.121312

[16] Fan, M., Hu, J., Cao, R., Ruan, W., Wei, X., 2018. A review on experimental design for pollutants removal in water treatment with the aid of artificial

intelligence. Chemosphere 200, 330-343.

[17] Akratos, C.S., J.N.E. Papaspyros, V.A. Tsihrintzis. An artificial neural network model and design equations for BOD and COD removal prediction in horizontal subsurface flow constructed wetlands, 2008

[18] Min H. (2010), Artificial intelligence in supply chain management: Theory and applications. International Journal of Logistics Research and Applications, 13 (1), p. 13-39.

[19] Davenport T.H., Ronanki R. (2018), Artificial Intelligence for the Real World: Don't start with moon shots, Harvard Business Review, 96 (1), p. 108-116.

Chapter 11

# Analyses of Open Security Issues for Smart Home and Sensor Network Based on Internet of Things

*Jung Tae (Steve) Kim*


## Abstract

A lot of communication are developed and advanced with different and heterogeneous communication techniques by integration of wireless and wire connection. Conventional technology is mainly focus on information technology based on computer techniques in the field of industry, manufacture and automation fields. It consists of individual skill and technique. As new technologies are developed and enhanced with conventional techniques, a lot of new application is emerged and merged with previous mechanism and skills. The representative application is internet of things services and applications. Internet of things is breakthrough technologies and one of the innovation industries which are called 4 generation industry revolution. Many different types of object and devices are embedded in sensor node. They are inter-connected with optimized open system interconnection protocol over internet, wireless and wire medium. Most of communication is fully inter-connected with conventional techniques at point to point and end to application in general. Most of information in internet of things is weak against attack. This may induce vulnerable features to unauthorized and outside attacker over internet protocol, Bluetooth, Wi-Fi, and so forth. As high and low efficient equipment are merged into heterogeneous infrastructure, IoT communication surroundings has become more complex, Due to limited resources in IoT such as small memory, low power and computing power, IoT devices are vulnerable and disclosed with security problems. In this chapter, we analyzed security challenges and threats based on smart home network under IoT service.


**Keywords:** Security issues, Vulnerability, Attack model, Internet of Things

## 1. Introduction

IoT (The Internet of Things) is widespread, ubiquitous and becoming realized in the real world. Recently, a lot of smart sensor nodes and objects are interconnected and co-operated via the Internet protocol. The Internet of things, its devices and objects are regarded as a global network infrastructure by linking physical and virtual objects. It is implemented by the merging of data capture and communication capabilities in sensor node. IoT is used for connecting devices and sensors with small and limited resources devices to detect a lot of different devices. These kinds

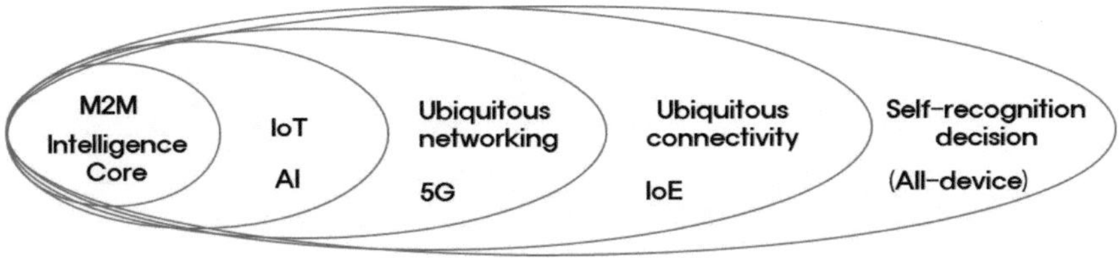


**Figure 1.**
*Evolution of trend of IoT based on development of technology.*

of infrastructures and connectivity include existing and involving embedded sensor networks via Internet and network [1]. Recently, advanced technologies in the semiconductor enable cost effective solutions to integrate wireless sensor network and connect application with embedded processors and sensors [2]. Previous works are focused on the security mechanism and data transmission in wire and wireless sensor network. Most of the Internet of Things consist of with RFID (Radio frequency identification), sensor devices, WSN (Wireless sensor network), internet and other network, etc. Information security issues are occurred during transmission. It becomes more complicated, critical and essential problems. With a number of things, objects, sensors and actuators which is connected to the Internet, a massive and real-time data flow can be automatically connected with different protocols. Most of papers are focus on in the field of efficient and reliable mechanism of security engines. It includes many applications such as sensing, privacy, tracking, services, data modeling and protocols. The main and issues are security field because the conventional security algorithm and mechanism are not used and suitable for its application with restricted resources. To apply enhanced security and privacy, constrained devices and light-weight cryptography is required for optimal security mechanisms. Therefore, cryptography mechanisms and security protocols should be optimized to adapt constrained devices and objects or new design method to be applicable and integrating into related IoT system. Many researchers enhance try to enhance the security mechanism and schemes. They also tend to improve and develop security protocol with high speed hardware regardless of limited condition [3]. Y. W. Lim et al. proposed reduced hardware architecture and system-on-chip targeting sensor node to achieve energy efficient on the IoT healthcare sensor node. They focused on reduced hardware architecture including sensor node's power consumption and cost [4]. In addition, lightweight cryptography make an alternative idea to implement light-weight security algorithm with low computational and small capabilities. In general, IoT system should be analyzed by its original requirements such as heterogeneous, resource constraints and dynamic environment. It provides its requirement in the field of network, cloud, user, attacker, platform and service. Evolution of IoT based on development of technology is shown in **Figure 1**. This paper is mainly summarized and contributed as follows. First, we introduce and discuss the concept. There is an overview and trend of the related works in the second section. In the third section, we provided requirement and consideration of security issues for IoT System. In the fourth section, open security issues of smart home network are analyzed. Lastly, we concluded in section five.

## 2. Related works

Security requirements for IoT application will be emphasized on the importance of formulated, implemented, and enforced security policies by their needs. Christof Paar et al. proposed described detailed analysis and method to be applicable for embedded security aspect concerning to IoT application. Regarding to traditional security solution, a lot of researcher and works have done and realized to provide embedded security system with small hardware resources such memory and low computation ability [5]. Jorge Granjal et al. surveyed and analyzed existing protocols and mechanisms for open research field. They analyzed that how existing approaches can be ensure fundamental security requirements and protect data in the field of IoT application. They also summarized the open challenges and strategies for the future work in this field [6]. Most security protocols which is used for network and internet security cannot be implemented with smart home systems related because they are low security complexity and vulnerable in smart home applications for the wireless sensor nodes. The major security issues for smart home systems are initial session key establishment between the wireless nodes and gateway or control box in smart home system. Yue Li proposed and analyzed a sort of lightweight key establishment protocol for home energy management systems. He presented an example of implementation and protocol in detail [7]. Kozlov et al. discussed about threats for privacy and security at a different architectural level of the smart home. They especially advertised to analyze privacy risk levels for privacy control mechanisms, methods, and energy aspects concerning to security, privacy, and trust. They are also evaluated an energy consumption in entire smart home infrastructure [8]. The security matters of information and network should be considered with representatives of properties such as identification, confidentiality, authentication, integrality and repudiation. In spite of a different requirement on Internet, IoT system will be applicable to the crucial and critical areas such as medical and health care, home, energy, intelligent transportation, smart factory and so on. Therefore, security needs in the IoT are more necessary and indispensable in availability and dependability. Generally speaking, the IoT can be divided into four layers [9]. The representative architecture and its configuration is described in **Figure 2** [10]. The structures of IoT are generally divided and classify into three layers. It consists of perception layer, application layer and network layer. Jeong Gi Lee et al. analyzed a current research and development trend. The main idea is that how integrated platform can be implemented with data security between different smart home nodes and devices. They implemented integration platform based on android to provide with simple development and scalability. It can be easy to access for authorized user. Smart home network based on related sensor products have a different ways of the data exchange. Its platform can be integrated easily and

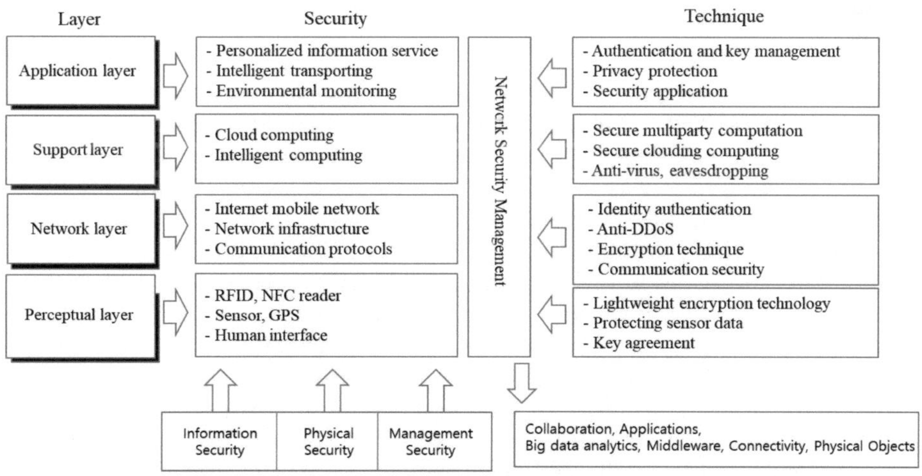


**Figure 2.**
*Basic architecture of IoT services.*

connected by heterogeneous network products and external transmission security processing for data communication. It can be supported to enable the integration of sensors [11]. Freddy K Santoso et al. proposed the implementation and design for IoT smart home system with embedded Wi-Fi system. It includes gateway to enable and activate secure communication between IoT sensor devices and control system. It allows user to control, access and configure related devices. It can be realized by smart phone and mobile devices to interface external communication and control devices [12]. Himanshu Gupta et al. presented a security framework for IoT applications using block chain technology and technique. It provided many unique characteristics such as better privacy, manageability, fault tolerance and scalability [13]. Musa G Samaila et al. proposed the IoT hardware platform security advisor. It provides three functionality features such as security requirement elicitation, security best practice guidelines. They also gave a guideline for lightweight cryptography design and implementation consideration. It includes a summary of the cryptographic algorithm [14]. Nikos Komninos et al. surveyed smart home security based on issues, challenges and countermeasures in smart grid application. They analyzed the most representative threats to smart grid environment and smart home system based on a lot of scenarios. They summarized a review of security countermeasures related to smart home as follows [15].

1. Confidentiality and privacy: Symmetric/Asymmetric encryption algorithms, zero knowledge proof systems and data obfuscation

2. Integrity: Cryptographic hashing techniques, digital watermarking, timestamps, session keys and sequence numbers.

3. Authenticity: Keyed cryptographic hash function, hash based authentication codes and MAC-attached messages

4. Non Repudiation: Mutual inspection with smart meters and unique keys for customer-AMI communication and AMI transaction logging

5. Availability: Alternative frequency channels according to hardcoded sequence, anomaly based IDSs and specification based IDSs

6. Authorization: Attribute based encryption, attribute certificates and attribute based access control system

Abdullahi Arabo analyzed cyber security challenges with the connected home ecosystem. He presented related background, motivation, development and demand for inter-connecting of different devices. The smart phone or mobile agent is used to provide a variety of function and capability to users [16]. Md. Mahmud Hossain et al. analyzed a detailed analysis of IoT system. It includes threat models, security issues, challenges and many attack models [17]. They provided a sort of open problems and issues in IoT security and privacy problems. This makes researchers to guide and solve the most critical and open problems. Pranay P. Gaikwad et al. surveyed these applications based on smart homes systems using internet of things [18]. They presented the problems and challenges which is occurred in IoT and smart homes application based on IoT system. Some solutions they proposed overcome some problems and challenges. We summarized the recent and breakthrough works in the field of security problems and privacy issues related to smart home application. A more extensive and detailed of related works have been published [19]. Monammed Ali Al-Garadi et al. surveyed of machine and deep learning (ML/DL) methods for IoT security. They also define thematic taxonomy of ML/DL for IoT security [20].

## 3. Requirement and consideration of security issues for IoT system

### 3.1 Basic concept of IoT

The domain of smart home environments is regarded and considered as a major factor and element for the future Internet. As a lot of homes are becoming smarter and smarter by using sensor and technology based on IoT, we can improve home security, energy efficiency, availability and comfortability. Consequently, to realize the future technology applicable to the smart home, we have to consider and treat with privacy into IoT environments. It can be identified and regarded as one of the major barriers and flaws. Because of the nature of the IoT environment, the appropriate security functions for secure and trustworthy smart home service would be applied extensively and considered importantly because the security threats will be increased and impact of security threats will be likely expanded. Jin-Hee Han et al. analyzed the requirement of security consideration for enhanced security and trustworthy mechanism in smart home system based on IoT environment [21]. As sensor nodes are widespread and utilized under ubiquitous environment, the security attacks on embedded device is increasing. The major factors include in the field of attacks such as crypto-analysis, physical, side channel, environmental, software and networks. Vijay Sivaraman et al. illustrated network-level security and privacy control for devices in smart home based on IoT. They proposed that software defined networking technology would be used to dynamically block and quarantine devices. It is based on their network activity [22]. The major security concerns for IoT system are summarized and included factors such as user identification, tamper resistant, secure S/W execution, secure content, secure data communications, identity management and secure storage. As a results of a risk and security analysis for a smart home automation system, it can be developed in collaboration with new schemes for leading industrial factors. They summarized the first steps and models of privacy and security for smart home applications. It is regarded as support and necessity for enforcing system security and user privacy, and it can help to realize the potential power in smart home environments. The typical architecture in IoT application can be divided and classify into three layers as following description [23]:

1. Perception Layer: In this layer, it collects, acquire and process the information from physical world. It is made up with two part to communicate with sensor devices and wire and wireless sensor network

2. Transmission Layer: In this layer, it transfers information in a large or long distance area. To connect and integrate information in perception layer, the information can transfer by using mobile, Wi-Fi and other communication media

3. Application Layer: In this area, it can process and service the information which is included in the layer.

Many applications provide middleware technology, computing technology and network processing in each layer. The main devices in perception layer include RFID, Zigbee and all kinds of sensors. Basic architecture of IoT service is shown **Figure 2**.

They are highly vulnerable to attacks. Several common attacks are included node capture, fake node and malicious data, denial of service attack, timing attack, routing threats, reply attack, side channel attack and mass node authentication problem. Network layer security problems have critical problems such as traditional security problem, compatibility problem, cluster security problems and privacy disclosure. In application layer, its security issues are different and more complex because of different industry or environment. The following elements should be solved with data access control, identification, data protection and recovery, authentication, ability of dealing with mass-data and software vulnerabilities in application layer. The IoT system has a particular restriction, constraints and limitation in terms of computational power, small memory and power. It makes significantly different from existing distributed systems. It can be recognized in real world that the existence of tiny computing devices is very much vulnerable to different security attacks as mentioned above. Security in level and requirement is shown in **Figure 3**.

Sye Loong Keoh et al. gave an overview of the efforts and demands in the IEFT (Internet Engineering Task Force) to standardize security solutions for IoT ecosystem. They provided a detailed review with communication security solutions for IoT. Especially, they used to conjunct with standard security protocols to be applied in the CoAP (Constrained Application Protocol), and application protocol to adapt the constraints IoT devices [24]. Pranay P. Gaikwad et al. presented the architecture of IoT related to attacks model. Smart home network can be operated with household devices and home appliances. It could monitor and control remotely with different connection and control ways. When these kinds of household devices in smart homes are connected with wire or wireless Internet under standard protocols. The whole system is so called as smart home network and can be realized in loT environment or smart homes based on IoT devices. They presented the problems and challenges which is occurred in loT and smart home applications. Some solutions that they proposed overcome and solve some problems and challenges in real solution matters [25]. The security design can be adapted with these kinds of diverse deployment scenarios. The representative ideas have a concise set of cryptographic, single security policy framework, security mechanisms, and configuration parameters with policy-dependent. These kinds of requirement and consideration in terms of system perspectives should take into account for entire system. In spite of IoT devices are constrained with limited resources, it can be deployed with easy steps and still has a vulnerability problem. Therefore, the traditional and conventional security mechanism and algorithms cannot be straightforward realized in smart things and sensor nodes. The major and representative limitation and constraints are shown in **Table 1** [26].

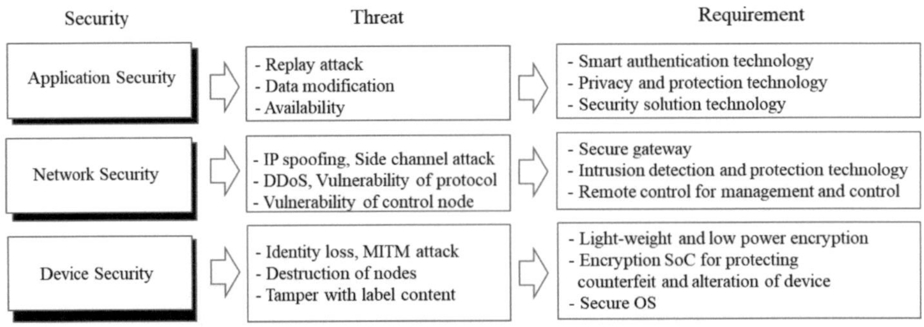


**Figure 3.**
*Level and requirement for security.*

| Hardware aspect | Software aspect | Network aspect |
|---|---|---|
| • Computational and energy constraints<br>• Memory constraint<br>• Tamper resistant | • Embedded software constraint<br>• Dynamic security patch | • Mobility, scalability, multiplicity<br>• Multiple medium of communication<br>• Multi-protocol networking<br>• Dynamic and stable network topology |

**Table 1.**
*Major security constraints of IoT devices.*

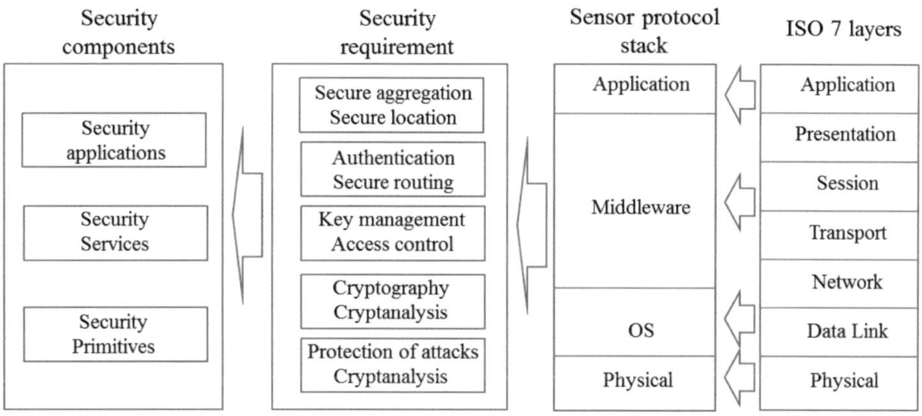


**Figure 4.**
*Basic component of security architecture.*

We analyzed the key element of security architecture with relation to sensor protocol, security demands and ISO7 layer as shown in **Figure 4**. There are many application layer protocol such as CoAP (Constrained Application Protocol), XMPP (Extensible Message and Presence Protocol) and MQTT (Message Que. Telemetry Transport), AMOP (Advance Message Queuing Protocol) [27]. Hee-jeong Kim and Jeong Nyeo Kim proposed end-to-end message security protocol based on ultra-weight cipher algorithm. This algorithm can increase security level and lower security overhead in resource limited communication [28].

### 3.2 Application of IoT services

The representative issues and services are included a lot of mechanisms. The main topics consist of end-to-end security, fault tolerance, key management, energy efficient security, trust management, IoT big data and it's forensic and so on. We also presented requirements of IoT System including basic principles as well as challenges and barriers.

a. Basic Principles

- Use standard and its application protocols.

- Detail protection and defense from malicious attacks.

- Secure algorithm is embedded in the system and realized with lightweight Algorithm.

- All code is implemented by authentic and trusted techniques.
- All protocols and communication is implemented by encrypted and authenticated techniques.
- All access and control to resources should be authenticated and authorized.

b. Challenges and Barriers

- Secure hardware platform is required and exploited.
- Complicated security design can be solved a cheap and mass production in silicon devices.
- A lot of vulnerable devices are revealed and plethora in the real world.
- Individual and collective risk can be monitored and assessed.
- Shared and distributed framework can be formulized and analyzed by decision making technique.
- Self-validating framework for monitoring and reasoning.

### 3.3 Security and privacy issues

The implementation of protocols with constrained networks should be dealt with some open problems. It is induced and related to the nature and feature of the physical devices. The features are included a limited computational capacity, a low amount of memory, and a limitation on energy computation, it makes the design of these protocols too hard and complicated in nature. We give some requirements and summarize for security and privacy issues related to IoT services as follows [29–30].

a. What we need to secure

- Access to sensors, devices and objects.
- Where the IoT network is located and inter-connected.
- What kind of data is generated and communicated.
- Whether the data which is produced is in active or at rest.
- A measured temperature is moderate to devices and sensors.
- Different secure complexity and level in each devices and objects.
- Suitable gateway system for multi-function is available.

b. Threat modeling for IoT

- Complex and large system is needed to guarantee and protect the attack model

- Unattended devices is produced and is exclusive.
- Public internet is connected and established.
- Many different threats should be considered.
- Broad and wide spectrum of countermeasures is required.

c. IoT Security Model

- Devices are small and scale down.
- Resources cannot be intensive and compact.
- A numbers of devices are required.
- Performance should be considered.
- Comprehensibility, manageability and availabilities.
- Different risk profiles are induced.
- One size does not fit all and complex and complicated techniques are necessary.

d. IoT Gateway

- Adaptation and extensible platform can be utilized.
- Rapid customization using adaptors can be easy implemented.
- Multi-vendor with different standards & protocols can support.
- Common gateway platform is essential and vital in secure.
- Proxy system for device management can be used.
- Gateway is made up with data and control channels, configuration, status monitoring, and device registration and inventory, etc.
- Security and access control is embedded in gateway engine.

One of the major problems related with IoT is the heterogeneous nature of devices. Shachar Siboni et al. analyzed several specific IoT testing scenarios based on different IoT devices [31]. Daewon Kim et al. present common security requirement for IoT device identification system. The requirements are more important when the identification information is used as the sensitive data such as authentication [32]. Franco Loi et al. developed a systematic and optimized method to identify and evaluate the security and privacy on various IoT devices. They categorize the threats along four dimensions such as confidentiality, integrity, access control and reflective attacks [33].

## 3.4 Model of attacks and threats based on IoT devices

There are a various of vulnerable attacks in sensor nodes, RFID and its application because of its restricted resources. Security threats and vulnerability to sensor protocols and nodes can be classified by strong and weak attacks. Weak attacks are practical and threats by observing and manipulating the communication channel to acquire the data between a server and device tags. Replay attacks and interleaving attacks are representative examples of weak attacks model. Strong attacks are feasible threats and means for an attacker to be compromised and acquired a data on target tag. A memory of sensor is very vulnerable to be compromised and have a small resource. It can be easily attacked by side channel effect because the low cost and capability tag is unlikely to be tamper-proof. The major strong attacks are included forward traceability, backward traceability, and server impersonation and described in Ref. [34].

It is possible to identify and realized with five distinct technology and trends in the future of IT, As many devices are widespread and ubiquitous with explosion. The future of IoT will be faced with representative characteristics as follows [35–37].

1. Data deluge: Amount of data is exploded and collected and exchanged through different networks. Forecasts indicate that more than several thousands of sensors and objects will be stored and integrated in the near future. Novel and unique techniques are needed to transmit, find and fetch the data in safe.

2. Miniaturization of devices: To realize the compact and small sensor, devices will be increasingly smaller and smaller.

3. Little energy consumption: the devices and system have to reorganize its own energy or self-organization to get a power.

4. Autonomic management: the devices or systems have to control its system and have a function for self-management and self-configuration capabilities.

5. IPv6 as an integration layer: It will provide nature network.

IoT application can be realized when we have a connectivity for anything from any time, any place connectivity for anyone. The enabling of technologies and realization of IoT application should be combined with RFID, NFC, sensor, smart technology and nano-technology. The characteristics of enabling technologies are summarized [9]. Internet of things' enablers should have characteristics as follows.

1. Energy: Energy harvesting technology and low-power chipsets are critical problem to develop and evolve IoT applications.

2. Intelligence: Devices should have a capability to reform and improve for self-organization and inter-machine communication, etc.

3. Communication: As the communication technology and means enables the devices to communicate with inter-networking. New materials and Integrated on-chip technology and smart multi-band frequency antennas are required.

4. Integration: Integration of smart devices into packaging and the products can be fabricated on-chip level.

5. It saves a significant cost and increase the applicable and friendly for the products and objects.

6. Interoperability: Protocols for inter-operability have to be customized for standardization.

7. Standards: Open standards mainly play an import role in the success of the IoT.

In general, energy-efficient communication standards, strategy of security and privacy should be considered into interest and needs. Compatible or identical protocols at the different frequencies are needed [30]. The function of sensor IoT is described in **Table 2**.

The hardware based issues related to sensors and objects should be dealt with in politics and laws. Enablers and objects of Internet of Things have a features as following description [38].

- Manufacturing, logistics and retail sectors: Next-generation industrial artificial intelligence, smart grid, supply chain, block chain, smart factory and inventory management, remoting control maintenance, anti-counterfeiting, and so on.

- Energy and utilities sectors: Smart inspection in electricity and energy, efficient energy and consumption mechanism, smart grid in transmission, real-time operation and monitoring system, smart grid in IoT connection and so forth.

- Artificial intelligent transportation support: Telematics, GPS and wireless networks, use of in-vehicle sensor networks, collaborative road safety and efficiency technology, vehicle tracking, developing smart vehicles and traffic data collection mechanism, etc.

- Environment monitoring systems: Remote inspection and control system for monitor the behavior on wireless sensor nodes such as environment condition, monitoring of weather, detection of soil condition, etc.

- Home management and monitoring: Smart sensor nodes, secure home gateway connected to security engines, electrical appliances, smart energy control, etc.

Example of IoT based on network topology is shown in **Figure 5**. To connect sensor networks with traditional communication networks to other networks, IoT

| Sensors | Function |
|---|---|
| RFID/NFC | To identify and track the sensor, object and so on. |
| Sensor | To collect and process the data flow<br>To detect the changes and moving in the physical status of things and objects |
| Smart technology | To enhance the power of the network<br>It is developing processing capabilities |
| Nano technology | To make the smaller and smaller things<br>The ability to connect and interact in small area. It can be fabricated on chip with CMOS process |

**Table 2.**
*Function of sensor IoT.*

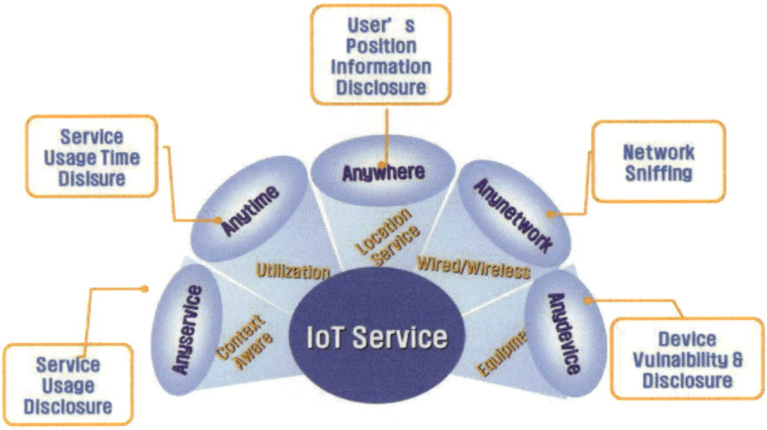


**Figure 5.**
*Application of IoT service [29].*

gateway can provide the function of protocol conversion and device management. As a general gateway, IoT gateway has characteristics and features such as access capability with a wide range, manageability for easy, interworking protocol for efficient topology [39]. Based on the home network application, secure gateway should provide and gather internal data to collaborate and aggregate in wireless sensor networks, and data transmission among Internet, high-speed, 5G networks, DSL networks, and other network interfaces. The system requirements of IoT gateway system should be considered by employing, protocol conversion, data forwarding and management and control [36]. Example of IoT system based on network topology is shown in **Figure 6**.

To analyze the performance and security analysis of IoT architecture, we can estimate the performance analysis such as computational cost, storage requirement and communication cost and security analysis. The security elements are included factors such as data confidentiality, tag anonymity, mutual authentication, data integrity, relay attacks and forward security. The representative attack and solutions are reviewed in **Figure 7**. We analyzed reusable security requirement and used them as examples of reusable security requirements. The attacks on IoT system are included denial of service attack, physical attack, tags cloning attacks,

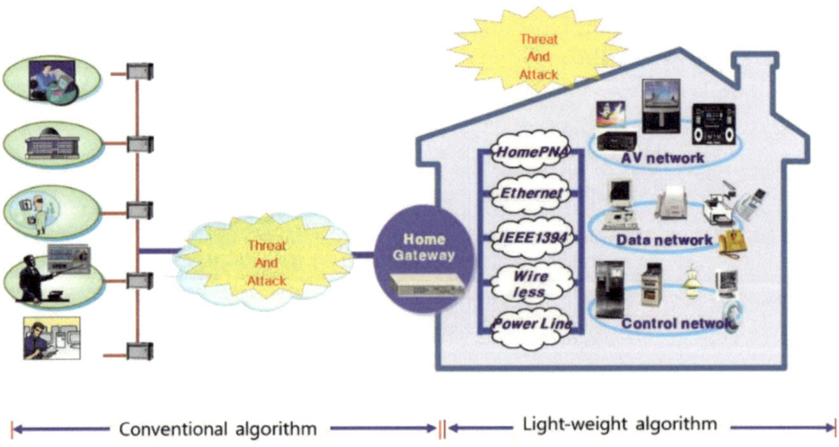


**Figure 6.**
*Example of IoT system based on network topology.*

| Security mechanism | Authentication method | Security service |
|---|---|---|
| Authentication | User's authentication | - PKI, OTP, certification of official recognition<br>- Home gateway authentication<br>- Fingerprint, vein |
| | Service authentication | Authentication between home service provider and home service |
| | Device authentication | - Authentication by certification<br>- Home gateway is operated as authentication server |
| Authorization Access control | User's access control | - Access control by policy of security level<br>- Access control profile by security level |
| | Device to device authentication | Necessity of access control between heterogeneous device |
| | Perception of situation Access control | - Restriction of user's service<br>- Access control by location based service |

**Figure 7.**
*Characteristics of authentication method and security mechanism.*

impersonation attack, replay attack and tags tracking. Pranay P. Gaikwad et al. presented open problems and challenges [40]. Denning et al. surveyed the security and privacy matters in IoT based smart homes, and provided a strategy for reasoning about security needs. They mentioned how to conduct an attack to avoid and have a feasibility in attack model. The attractive point they propose is that the system compromise a platform and attack model caused by damage [41]. In generally speaking, smart home system can be organized and configured as shown in **Figure 7**. Smart home system consists of home gateway, home server, and smart home sensor and devices. To enhance security complexity, we should take into consideration basic security functions for the smart home system. Security functions are against security vulnerabilities and security flaws caused by device resources issues, attacks, etc. The necessity and requirement of security in authentication level can be classified by authentication and authorization level. The representative characteristics and features are shown in **Figure 7**.

Computing service with confidentiality and threat factors under ubiquitous surrounding should be regarded as vulnerable attacks and insecure services viewpoints. The major factors to be considered are as follows [42].

- Confidentiality for secure data
- Authentication and access control for authorized a user
- Integrity and non-repudiation for reliable data
- Availability and survivability for fusibility
- Privacy and authority control for authorization

Also, threat factors and features can be considered as follows.

- Sensor node attack

- Eavesdropping
- Sensing data privacy
- DOS (Denial of service)

A brief of characteristics and features of smart home network is depicted as follows.

- Home network may not be isolated in sub-network and inter-connected for operability.
- Wi-Fi and PLC signals may propagate and be widespread to the next door. It can be exposed.
- Different network and addressing scheme are required.
- IP address, group and node ID is required in separated.
- Many devices have a low computational power and small memory capability.
- No public key cryptography is required. It uses a symmetry algorithm.

To enhance advanced and high complexity security, we have to take into account for intrusion protection and detection mechanism in sensor network. Especially, home gateway should be required enhanced security framework, authorized control, privacy and resilience of survival, etc. The security issues with heterogeneous connection between devices under IoT application cannot be solved with conventional cryptography techniques. Many works have done and cannot utilize existing PKI (Public key infrastructure) and lightweight schemes. Some researchers have evolved new scheme with ultra-lightweight algorithm. It has still unsolved problems. Many researcher and works tend to focus on following techniques relating to smart home network security.

- Integration of security on middleware is realized.
- New security function and schemes beyond security function of middleware are utilized.
- Safety about middleware security function can be analyzed.
- Lightweight algorithm is developed and can be suitable for its system.
- User's convenience can be provided with simple algorithm

## 4. Open security issues for smart home network

We analyzed and discussed the critical and essential issues for open solution of IoT related to smart home network. It mainly focuses on challenges and mechanism in application field. Example of IoT based network topology and attack model on home network is shown in **Figure 8**. We can estimate the security requirements for smart home network. It consists of device authentication, network monitoring,

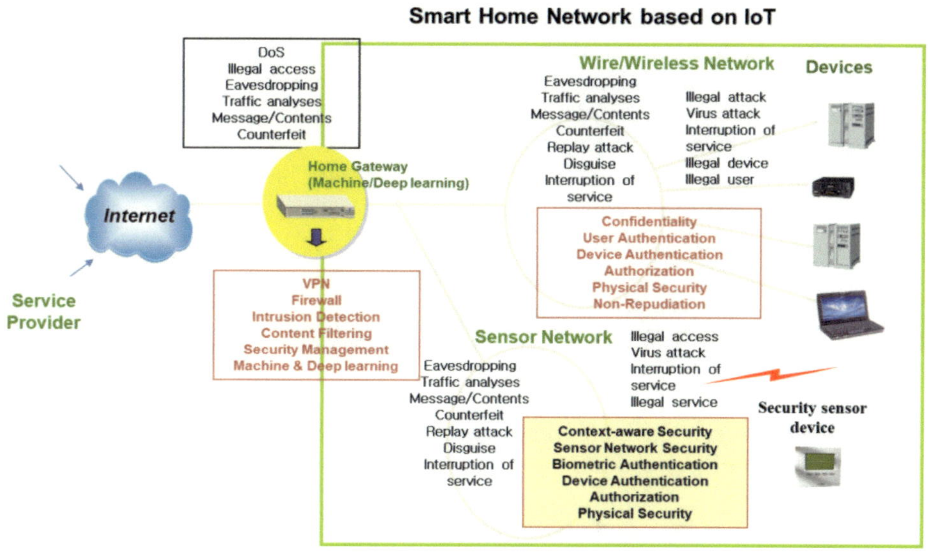


**Figure 8.**
*Example of topology for attack model on smart home network.*

secure session key management, physical protection, information security, and user authentication [43]. Andreas Jacobsson et al. reviewed a risk analysis related to smart home automation and control system [23].

The reviewed and related works presented have included risk analysis based approaches, security based approaches, privacy based approaches and industry based approaches. Chakib Bekara analyzed security matters and challenges for the IoT based on smart grid [44]. He considered security issues as a cyber and physical system, the IoT based smart grid will be faced with several security issues such as impersonation, eavesdropping, data tampering, authorization, control access, privacy issue, malicious code, availability and DoS issues. When dealing with security algorithms, several challenges should be considered with scalability, mobility, deployment, legacy systems, constraints resources, heterogeneity, interoperability, bootstrapping, trust management and latency and time constraints. Noy Hadar et al. proposed an innovative and creative cloud-based framework to protect attacks and threats on IoT devices. It can be applicable to cost effective solution [45].

| Sensors | Function |
|---|---|
| Device security | Insecurity due to device category and capability |
| | Software and firmware security |
| | Storage security |
| Communication security | Security service security |
| | Network service security |
| | Cryptographic security |
| Security service | Native service security |
| | Cloud service security |
| | Partner cloud service security |

**Table 3.**
*Critical security weakness of IoT system.*

To understand IoT security issues, we should investigate the characteristics of component for the IoT sensor and network among them. It consists of sensors in home network, sensor bridge network such as home gateway and cloud. Md. Mahmud Hossain et al. analyzed critical security weakness [17]. The security characteristics of sensors are shown in **Table 3** [46].

The security challenge of devices and sensors should be solved and included as follows. We analyzed the following issues.

a. Devices are not connected and reachable

   Most of the time, device is not connected

b. Devices can be stolen and lost

   It makes security difficult and uncountable when the device is not attached and connected

c. Devices are not included crypto-engines

   Strong security is difficult to solve the problem without high-speed processing power

d. Devices have finite life

   Devices have small power and needs for self-organization and recharge mechanism

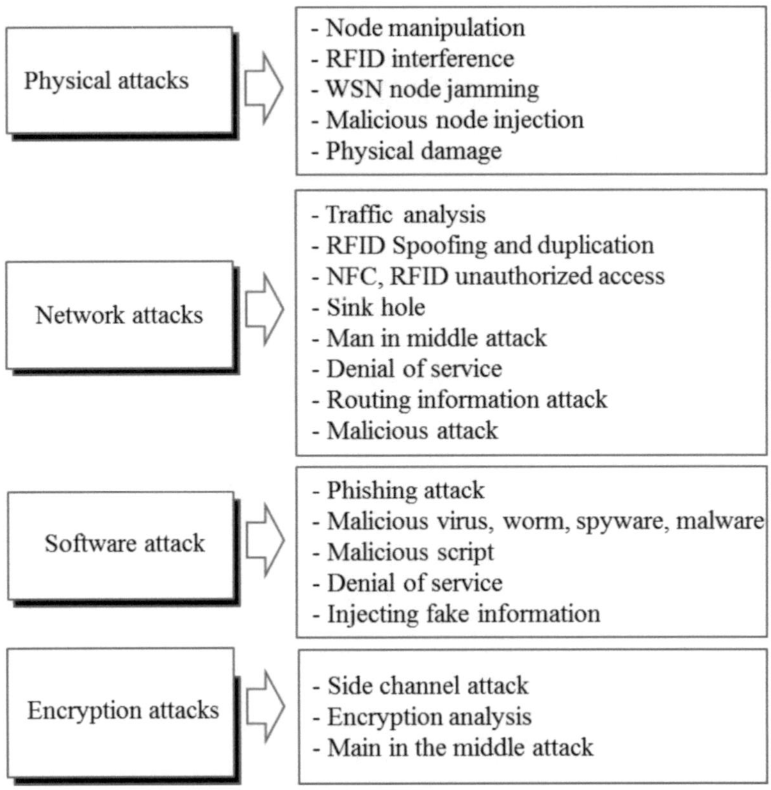


**Figure 9.**
*Attacks model and its characteristics.*

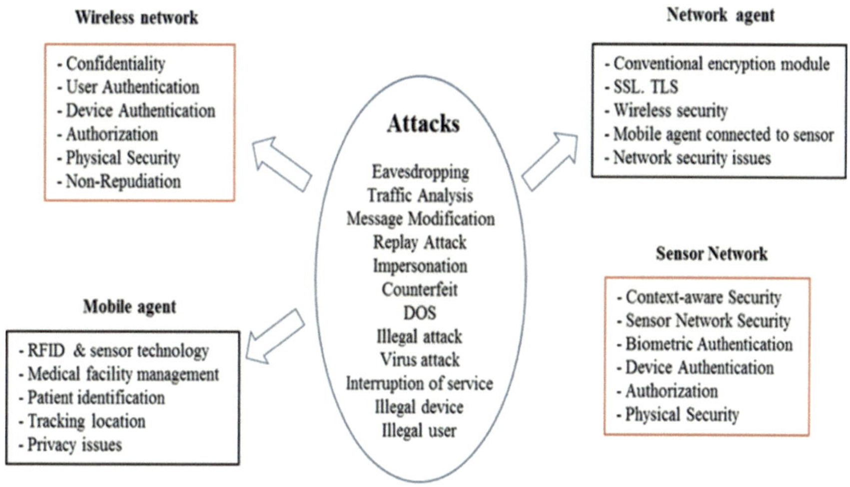


**Figure 10.**
*A variety of attacks on smart home network.*

e. Devices are transportable

Will cross borders

f. Devices can be recognized and detected by many readers

What data is exposed and disclosed to reader?

Characteristics of attacks and model are shown in **Figure 8**. The gateway can provide secure communication for authentication, secure session key exchange and monitoring between devices and gateway [47]. The attacks model and its characteristics are shown in **Figure 9**.

We also described the challenges of open issues for IoT's development. Trade-off between hardware, software and co-design method is needed for security framework and architecture for IoT system. We have to focus on delivering the creative and unique solutions for the state-of-the-art and innovative mechanism. It includes IoT security controls, optimized mechanisms for the new skills and extremely complex embedded applications. Example of IoT based on smart home network topology is shown in **Figure 8**. Example of attack model on home network is analyzed in literature [48]. Recent DDos (Distributed denial of service) attack and technology are developing with recent techniques and cause a high vulnerability on IoT system. It focuses on the requirement of scalable and optimized security solution. Ellia and Nayeem described common internet things and vulnerabilities [49]. A variety of attacks on smart home network is shown in **Figure 10**.

## 5. Conclusion

Internet of Things is used in many applications in different areas. loT has been already designed for industrial wireless sensor network in many applications. It can be developed for smart homes system in the near future. Architecture of loT and smart homes based on loT are analyzed in this review paper. In spites of many opportunities and recent technologies, many challenges and issues are produced by a lot of attacks which is addressed in IoT. It will inherit the drawbacks of the internet used in nowadays. We discussed about security mechanisms, we also point out that the challenges of open issues for IoT's development. Hardware and software co-design methodology is needed to fit into security framework and architecture for IoT system. We have to focus on delivering the current state-of-the-art IT security controls, optimized mechanisms for Internet of things and objects in the future works.

## Conflict of interest

"The authors declare no conflict of interest."


## Author details

Jung Tae (Steve) Kim
Mokwon University, Daejeon, South Korea

*Address all correspondence to: jtkim5068@gmail.com

## References

[1] Brahim Ethem Bagci, Shahid Raza and Tony Chung, Combined Secure Storage an Communication for the Internet of Things, In Proceedings of the 2013 IEEE International Conference on Sensing, Communications and Networking, 2013, pp. 523-531.

[2] Freddy K Santoso and Nicholas C H Vun, Securing IoT for Smart Home System, In Proceedings of the 2015 IEEE International Symposium on Consumer Electronics, 2015, pp. 1-2.

[3] Raja Benabdessalem, Mohamed Hamdi and Tai-Hoon Kim, A Survey on Security Models, Techniques and Tools for the Internet of Things, In Proceedings of the 2014 7th International Conference on Advanced Software Engineering & Its Application, 2014, pp. 44-48.

[4] Y. W. Lim, S.B. Daas, S. J. Hashim, R. M. Sidek, N. A. Kamsani and F. Z. Rokhani, Reduced Hardware Architecture for Energy Efficient IoT Healthcare Sensor Nodes, In Proceedings of the 2015 IEEE International Circuits and Systems Symposium, 2015, pp. 90-95.

[5] Christof Paar and Andre Weimerskirch, Embedded Security in a Pervasive World, Information Security Technical Report, Elsevier, Vol 12, Issue 3, 2007, pp.155-161.

[6] Jorge Granjal, Edmundo Monteiro, and Jorge Sá Silva, Security for the Internet of Things: A Survey of Existing Protocols and Open Research Issues, IEEE Communication Surveys & Tutorials, 2015, Vol.17, No.3, pp.1294-1312.

[7] Yue Li, Design of a Key Establishment Protocol for Smart Home Energy Management System, In Proceedings of the 2013 Fifth International Conference on Computational Intelligence, Communication Systems and Networks, 2013, pp.88-pp.93.

[8] D. Kozlov, J. Veijalainen, and Y. Ali, Security and Privacy Threats in IoT Architectures, In Proceedings of the 7th International Conference on Body Area Networks, 2012, pp.100-102.

[9] Zhibo pang and Junzhe Tian, Ecosystem-driven Design of In-home Terminals Based on Open Platform for the Internet-of Things, International Transactions on Advanced Communications Technology, 2014, Vol.3, Issues 1, pp.369-377.

[10] Kai Zhao and Lina Ge, A Survey on the Internet of things security, In Proceedings of the 2013 9th International Conference on Computational Intelligence and Security, 2013, pp.663-667.

[11] Jeong-Gi Lee, Chul-Seung Yang, Hun-Ha Kim and Kang-Jin Kim, A Research and Development of Integrated Platform for Data Security between Different Smart Home Devices, Journal of the Korea Institute of Information and Communication Engineering, 2015, Vol.19, No.5, pp.1173-1179.

[12] Freddy K Santoso and Nicholas C H Vum, Securing IoT for Smart Home System, In Proceedings of the 2015 IEEE International Symposium on Consumer Electronic, 2015, pp.11-12.

[13] Himanshu Gupta and Garima Varshney, A Security Framework for IoT Devices against Wireless Threats, 2017 2$^{nd}$ International Conference on Telecommunication and Networks, 2017, pp.100-106.

[14] Musa G. Samaila, Joao b. f. Sequeiros, Tiago Simoes, Mario M. Freire and Pedro R. M. Inacio, IoT-

HarPSecA: A Framework and Roadmap for Secure Design and Development of Devices and Application in the IoT Space, IEEE Access, Jan, 2020, pp.16462-16494.

[15] Nikos Komninos, Eleni Philippou and Antreas Pitsillides, Survey in Smart Grid and Smart Home Security: Issues, Challenges and Countermeasures, IEEE Transaction Survey & Tutorials, 2014, Vol.16, No.4, pp.1933-1954.

[16] Abdullahi Arabo, Cyber Security Challenges within the Connected Home Ecosystem Futures, Procedia Computer Science, 2015, pp.227-232.

[17] Md. Mahmud Hossain, Maziar Fotouhi and Ragib Hasan, Towards an Analysis of Security Issues, Challenges, and Open Problems in the Internet of Things, In Proceedings of the 2015 IEEE World Congress on Services, 2015, pp.21-28.

[18] Pranay P. Gaikwad, Jyotsna P. Gabhance and Snehal S. Golait, A Survey Based on Smart Homes System Using Internet-of Things, In Proceedings of the 2015 International Conference Computation of Power, Energy, Information and Communication, 2015, pp.330-335.

[19] Romano Fantacci, Tommaso Pecorella, Roberto Viti and Camillo Carlini, Short Paper: Overcoming IoT Fragmentation through Standard Gateway Architecture, In Proceedings of the 2014 IEEE World Forum on Internet of Things (WF-IoT), 2014, pp.181-182.

[20] Mohammed Ali-Garadi, Amr Mohamed, and Abdulla Khalid Al-Ali, A Survey of Machine and Deep Learning Methods for Internet of Thing Security", IEEE Communications Surveys & Tutorials, Vol.22, No.3, 2020. pp.1646-1685.

[21] Jin-Hee Han, YoungSung Jeon and JeongNyeo Kim, Security Considerations for Secure and Trustworthy Smart Home System in the IoT Environment, ICTC 2015, pp.1116-1118.

[22] Vijay Sivaraman, Hassan Habibi Gharakheili, Arun Vishwanath, Roksana Boreli and Olivier Mehani, Network-level Security and Privacy Control for Smart-home IoT Devices, In Proceedings of the 8th International Workshop on Selected Topics in Mobile and Wireless Computing, 2015, pp.163-167.

[23] Andreas Jacobsson, Martin Boldt and Bengt Carlsson, A Risk Analysis of a Smart Home Automation System, Future Generation Computer Systems, 56, 2016, 719–733.

[24] Sye Loong Keoh, Sandeep S. Kumar and Hannes Tschofenig, Securing the Internet of Things: A Standardization Perspective, IEEE Internet of Things Journal, 2014, Vol.1, No.3, pp.265-275.

[25] Pranay P. Gaikwad, Jyotsna P. Gabhane and Snehal S. Golait, A Survey Based on Smart Homes System Using Internet-of Things, In Proceedings of the 2015 International Conference on Computation of Power, Energy, Information and Communication, 2015, pp.330-335.

[26] Hailong Feng and Wenxiu Fu, Study of Recent Development about Privacy and Security of the Internet of Things, In Proceedings of the 2010 International Conference on Web Information Systems and Mining, 2010, pp.91-95.

[27] Sowmya Nagasimaha Swamy, Dipti Jadhav and Nikita Kulkarni, Security Threats in the Application Layer in IoT Application", International Conference on I-SMAC(IoT in Social, Mobile, Analytics and Cloud, 2017, pp.477-480.

[28] Hee-jeong Kim and Jeong Nyeo Kim, "A study of end-to-end message security protocol based on lightweight

ciphers for smart IoT devices, Journal of the Korea Institute of Information Security & Cryptology", Vol.28, No.6, 2018, pp.1309-1317.

[29] Shivangi Vasho, Jyotsnamayee Ram and Janit Modi, Internet of Things (A Vision, Architectural Elements, and Security issues), International Conference on I-SMAC(IoT in Social, Mobile, Analytics and Cloud), 2017, pp.492-496.

[30] Jung Tae Kim, Requirement of IoT Security Gateway to Improve Vulnerability of Sensor Node, 2015 International Conference on Future Information & Communication Engineering, Vol.7, No.1, 2015, pp.435-428.

[31] Shachar Siboni, Vinay Sachidananda, Yair Meidan, Michael Bohadana, Yael Mathov, Suhas Bhairav, Asaf Shabtai and Yuval Elovici, Security Tested for Internet of Things Devices", IEEE Transaction on Reliabilities, Vol.68, No.1, March 2019, pp.23-44.

[32] Daewon Kim, Jeongnyeo Kim, and Yongsung Jeon, Common Security Requirement for Device Identification System in Internet of Things, Advanced Science and Technology Letters, Vol.139, 2016, pp.306-309.

[33] Franco Loi, Arunan Sivanathan, Hassan Habibi Gharakheili, Adam Radford and Vijay Sivaraman, Systematically Evaluating Security and Privacy for Consumer IoT Devices, Proceedings of the 2017 workshop on IoT Security and Privacy, 2017, pp.1-6.

[34] Mouza Bani Shemaili, Chan Yeob Yeun, Khalid Mubarak and Mohamed Jamal Zemerly, A New Lightweight Hybrid Cryptographic Algorithm for the Internet of Things, The 7th International Conference for Internet Technology and Secured Transactions, 2012, pp.87-91.

[35] Jung Tae Kim and Hakkee Jung, Analyses of Attacks Model and Requirement for Smart Home based on Internet of Things, ICFICE2017, pp.100-102.

[36] Jung Tae Kim, Analyses of Secure Authentication Scheme for Smart Home System based on Internet of Things, IEEE Internal Conference on Applied System Innovation, 2017, pp.100-104.

[37] Jung Tae Kim, Analyses and Security Requirements for Smart Home Network based on Internet of Things, The 11th 2016 Internal Interdisciplinary Workshop Series, 2016. pp.100-104.

[38] Azzedine Boukerche and Yonglin Ren, A Secure Mobile Healthcare System Using Trust-based Multicast Scheme, IEEE Journal on Selected Areas in Communications, Vol.27, No.4, 2009, pp.387-397.

[39] H.Chan, A.Perrig, Security and Privacy in Sensor Networks, IEEE computer, Vol.36, No.10, 2013, pp.103-105.

[40] Pranay P. Gaikwad, Jyotsna P. Gabhane and Snehal S. Golait, A Survey based on Smart Homes System Using Internet of Things, In Proceedings of the 2015 International Conference on Computation of Power, Energy, Information and Communication, 2015, pp.330-335.

[41] T. Denning, T. Kohno and H. M. Levy, Computer Security and Modern Home Communication of ACM, 2013, Vol.56, No.1, pp.94-103.

[42] Jung Tae Kim, Analyses of Vulnerability for Healthcare System on Internet of Things, The 3rd Asia Workshop on IT Convergence of KIICE2017, 2017, pp.87-88.

[43] Changmin Lee, Luca Zappaterra, Kwanghee Choi and Hyeong Ah Choi, Securing Smart Home: Technologies,

Security Challenges, and Security Requirements, Workshop on Security and Privacy in Machine-to-Machine Communications, 2014, pp.67-72.

[44] Chakib Bekara, Security Issues and Challenges for IoT-based Smart Grid, Procedia Computer Science, 34, 2014, pp.532-537.

[45] Noy Hadar, Shachar Siboni, and Yuval Elovici, A Lightweight Vulnerability Mitigation Framework for IoT Devices, Proceedings of the 2017 Workshop on IoT Security and Privacy, 2017, pp.71-75.

[46] Jung Tae Kim, Analyses of Requirement for Secure IoT Gateway and Assessment, Information, Vol.19, No.3, 2016, pp.833-840.

[47] Jorge Granjal, Edmundo Monterio and Jorge Sa Silva, Security for the Internet of Things: A Survey of Existing protocols and Open Research Issues, IEEE Communication Survey & Tutorials, Vol.17, No.3, 2015, pp.1294-1312.

[48] Jung Tae Kim, "Analyses of Security Issues for Smart Home System based on Internet of Things", International Conference on Applied System Innovation, 2016, pp.160-1645.

[49] Elisa Bertino and Nayeem Islam, "Botnets and Internet of Things Security", IEEE Magazines, Computer, Feb, 2017, pp.76-79.

Chapter 12

# IIoT Machine Health Monitoring Models for Education and Training

*Ishwar Singh, Sean Hodgins, Anoop Gadhrri and Reiner Schmidt*

## Abstract

IoT, IIoT and Industry 4.0 technologies are leading the way for digital transformation in manufacturing, healthcare, transportation, energy, retail, cities, supply chain, agriculture, buildings, and other sectors. Machine health monitoring and predictive maintenance of rotating machines is an innovative IIoT use case in the manufacturing and energy sectors. This chapter covers how machine health monitoring can be implemented using advanced sensor technology as a basis for predictive maintenance in rotating devices. It also covers how sensor data can be collected from the devices at the edge, preprocessed in a microcontroller/edge node, and sent to the cloud or local server for advanced data intelligence. In addition, this chapter describes the design and operation of three innovative models for education and training supporting the accelerated adoption of these technologies in industry sectors.

**Keywords:** IoT and IIoT, Machine Health Monitoring, Neural Network, Predictive Maintenance, Wireless Monitoring

## 1. Introduction

To improve safety and performance, manufacturing companies have been considering the adoption of advanced technologies, as stipulated in Industry 4.0 reports. These technologies include IoT/IIoT, big data and analytics, smart factory, cyber-physical systems (CPS), and interoperability of automation equipment. Over the last few years Artificial Intelligence (AI) has comeback with a vengeance not seen before in any of modern technology implementations. AI and IoT/IIoT are driving forces in Industry 4.0 with applications in manufacturing, oil and gas, utilities, banking, aerospace and defense, healthcare, retail, telecommunications, smart cities, and transportation. Artificial intelligence's impact on manufacturing can be organized into the following main areas: product quality and yield, predictive maintenance, collaborative robots, generative design, supply chain management and safer work environment.

While there are a lot of papers written about predictive maintenance and AI applications [1–7], but there is a lack of machine health predictive maintenance teaching and training models for university level courses as well as facilities for providing hand-on interactive experiences in the use of IoT/IIoT and Industry 4.0 platforms. To address this need SEPT created a learning factory and a framework for learning these technologies as well as designed and developed machine health monitoring models.

To reinforce hands-on learning two key aspects of machine health monitoring are presented in this chapter: foundational technologies such as sensors for predictive maintenance, IoT and IIoT ecosystem, and CPS monitoring tools; the second aspect covers in detail the design, development, and implementation of machine health learning models along with implementation of AI tools for real-time data generation, preprocessing it and sending to the cloud or local server for data analytics and visualization. These models provide an opportunity for developing and learning multidisciplinary and multi-capability skills in a laboratory setting.

## 2. SEPT learning factory

The engineering education is witnessing revolutionary changes in response to the huge demand for engineers with high industry 4.0 competencies. Engineering graduates will need to learn IoT and IIoT as the foundation for implementing Industry 4.0 concepts in industrial operations. The school of Engineering Practice and Technology (SEPT) at McMaster University has recently made huge effort to integrate IoT, IIoT and Industry 4.0 in the undergraduate and graduate curriculum [8].

In the undergraduate Automation Engineering Bachelor of Technology offered by SEPT, a new smart systems specialization is introduced in the fourth year, where the offered courses focus on IoT. The other option is the Industrial Automation specialization. A new introductory IoT course (SMRTTECH 3CC3) was developed and offered for the first time at the 3A level in the fall of 2019. The purpose of this course is to introduce the students to the fascinating world of IoT before choosing their specialization for the fourth year and before going to their mandatory co-op training [9].

Another effort by SEPT is the formation of a Cyber-Physical Systems Learning Centre that focuses on implementing Industry 4.0 concepts for teaching, training, and research at McMaster University [10, 11]. The Centre includes a series of specialized learning labs and the SEPT Learning Factory that allow the development of various theoretical and technical skills needed for product production. The Learning Centre complements students' qualifications and abilities by providing new technical skills that emphasize the inherent multidisciplinary nature of smart systems and advanced manufacturing.

### 2.1 Machine health monitoring

Machine health monitoring is a key opportunity to improving and maintaining profit. In manufacturing industries, it is expected that when a machine is started it should perform as designed and used for an application and run for hours, months and years without any breakdown interruptions. Vibration monitoring along with power and sound monitoring are a significant source of machine health information. By adopting the use of new technologies such as IIoT can lead to improved manufacturing productivity with more reliability and even reduce skill requirement needs. However, developing and implementing these applications does require a new breed of engineering technology graduates. IIoT offers an opportunity for ubiquitous detection of machinery faults that can lead to prescriptive maintenance plans.

During operation mechanical faults in machines produce unique vibrations which depend upon the geometry of the machine elements such as shaft, spindle etc., and shaft rotation speed, in addition to the obvious load factor. There is a huge list of mechanical faults that can be detected with vibration data collection and performing an analysis on it. This list includes imbalance; misalignment; bent shaft; rubbing shaft; bearing defects; loose parts; and belt drive faults etc. For example, recently a pump servicing company has identified a few major causes of vibration

[12]. Their report lists six main causes of pump vibration problems and anyone of those could take a pump out of service for unplanned and expensive repairs. A pump's poor performance can be due to one of the following vibration problems: pump cavitation; bent pump shaft; pump flow pulsation; pump impeller imbalance; pump bearing issues; and misalignment of the shaft.

### 2.2 IoT and IIoT implementations

Many organizations use manual and off-line inspection tools for their reliability and maintainability programs while IIoT provides an opportunity to perform these tasks on-line in real-time time. IoT alone is projected to deliver between $1.9 and $4.7 trillion of economic value by 2025. The IIoT for asset monitoring is expected to produce $200-$500 billion in economic value by 2025 [13]. These technologies enable machine health monitoring (or also referred as condition monitoring) and predictive maintenance to optimize maintenance processes and improve operating costs. This type of monitoring is expected to help manufacturers to optimize their operating costs by predicting the failure of critical machines and their components to achieve high efficiency and reliability.

According to one market report the global machine health monitoring market size, driven by on-premises deployment, is estimated to reach USD 3.9 billion by 2025 [14]. On-premises application development give organizations control over their data and systems to protect the critical information. Whereas deployment on a remote cloud has its advantages and disadvantages related to hardware, software, deployment, and maintenance costs. Another factor that needs to be considered carefully is storage capacity at the local and remote server sites. Cloud-based deployments do provide organizations with enhanced accessibility and scalability, 24/7 service speed, and IT security measures that cannot be implemented due to lack of resources at the local site. In part the growth of this market is being driven by the availability of secure cloud platforms. An overall IoT/IIoT ecosystem with elements shown in **Figure 1** would be required. This type of ecosystem has been established at the SEPT Learning Factory to demonstrate these technologies [15].

Online machine health monitoring systems can be implemented for critical equipment, such as motors, turbines, blowers, pumps, and compressors, that have an immediate impact on the productivity of plants as well as human and machine safety, and the environment. Current monitoring systems include a sequence of sensors permanently mounted on the critical machines for sensing. The sensors are connected to microcontrollers, single board computers, and/or PLCs, and generated data can be sent to a central server of a plant or to an outside cloud platform such as

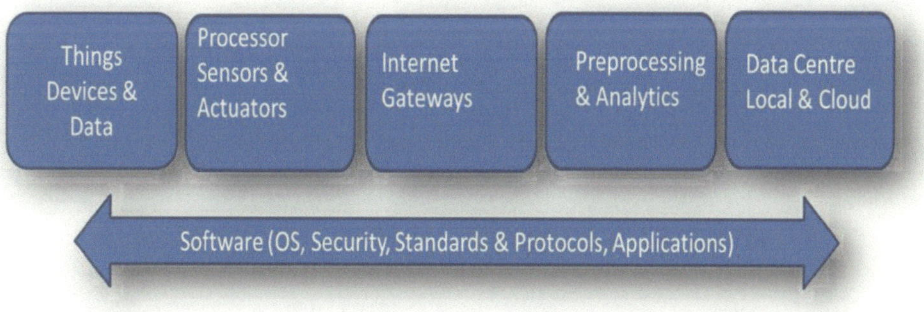


**Figure 1.**
*IoT/IIoT ecosystem overview.*

PubNub, Google, Amazon, and ThingsBoard etc. The sensor data is sent to the plant operators through either a wireless network or a cabled network and displayed on monitors. This can be accomplished in two manners as depicted in **Figures 2** and **3**. UniS is the UniSphere 1 local platform at SEPT hosting MQTT broker, **Figure 4**, as well DDS server.

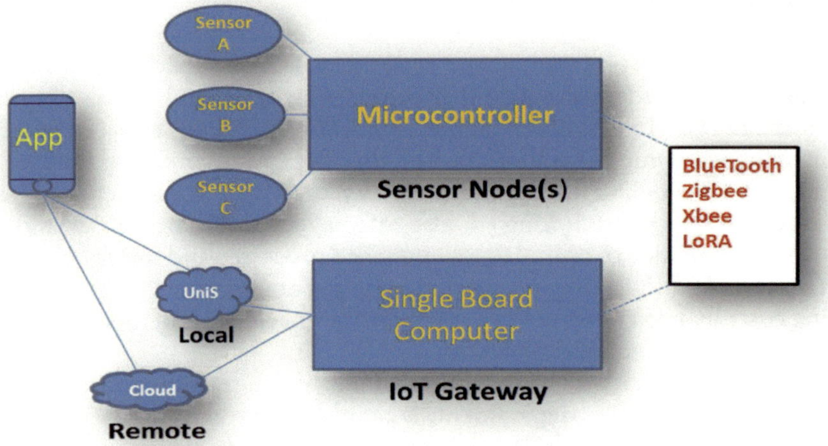


**Figure 2.**
*A typical IoT model used in SEPT learning factory.*

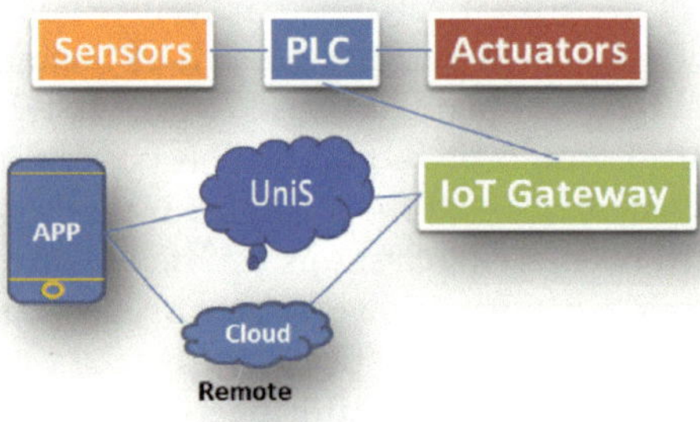


**Figure 3.**
*An IIoT model used in SEPT learning factory.*

**Figure 4.**
*MQTT broker local and remote servers.*

## 3. Sensors used for predictive maintenance

Vibration, sound pressure, motor current, magnetic field, temperature, and oil quality are some of the more common sensors used for condition-based monitoring for predictive maintenance tasks. Most systems will only employ some of these sensors based on potential critical faults detectable by the selected sensor. Vibration sensors fall in this category.

### 3.1 Accelerometer sensor

This type of sensor is used for general purpose applications for vibration and shock measurements. They are available as digital or analog devices and designed using different principles such as piezoelectric, piezoresistive and capacitive. These components are used to convert the mechanical motion caused in the device into an electrical signal. Accelerometers are the most used vibration sensors for predictive maintenance of turbines, pumps, motors, and gearboxes. The piezo vibration sensors are considered as the gold standard in this category. An important category for acceleration measurements are the MEMS sensors. MEMS accelerometers and microphones are highly suited to battery-powered predictive maintenance systems due to their small size, low power consumption, and high-performance capabilities [16].

### 3.2 Strain gage sensor

Strain gage is a sensor whose resistance changes with an applied force. From resistance change of the *strain gauges* a voltage change can be obtained with an electrical conditioning circuit such a voltage divider or Wheatstone bridge. This voltage is *measured* and is related to the *vibration* components. The amplitude and frequency of the *vibration* are changing during a machine's operation. These sensors have the advantage over many other types that they can be used for curved surfaces.

### 3.3 Velocity sensor

Velocity sensors are used to measure a frequency range of 1 – 1000 Hz. The sensors are suitable for vibration monitoring and balancing applications on rotating machinery. This type of sensor has a lower sensitivity for vibrations with high frequencies than accelerometers and is therefore less susceptible to overload. These sensors are used for high-temperature applications like above 700°F.

### 3.4 Gyro sensor

The main functions of the gyro sensor for various applications are angular velocity sensing, angle sensing, and control mechanisms. Gyroscope sensors are used in the car navigation systems, electronic stability control systems of vehicles, drones and radio-controlled helicopters and robotic systems. There are different types of gyros for different applications and they are: ring laser; fiber-optic; fluid gyro; and vibration gyros. Most used, vibration gyro sensors, use either piezoelectric or silicon transducer element in their construction. Gyro sensos, when used in conjunction with accelerometers, can keep track of the orientation of the system, thus providing a complete picture of the vibrating system.

### 3.5 Non-contact sensors

Most used non-contact vibration sensors types fall in the following categories: microphone or acoustic pressure sensor; laser displacement sensor; and eddy current or capacitive displacement sensor. Non-contact vibration sensors can be deployed on both new and old machines, even when they are hot, wet, in a hard-to-reach places or too small for other sensor types.

## 4. Predictive maintenance and AI

Predictive maintenance as evolved over the years now is considered as an important IIoT application that can be implemented fairly-easily. It also connects with Industry 4.0 paradigm, big data and analytics, and machine learning. Over the years it has evolved from reactive, preventive, and reliability-based maintenance operations [17]. Many new online sensors are being introduced each year. IIoT enables machine condition control in the following manner: collection of real-time sensor data, perform data analytics, followed by corrective action either by an operator or autonomously by the machine. In this context data analytics is related to converting data into actionable information and can have implications for predicting future events such as predictive maintenance. Now with the revised promise of AI technology the predictive maintenance can evolve to prescriptive maintenance where a smart machine can help avoid the predicted failure. In essence the following steps are required to implement such program: identifying critical assets, creating a database of relevant information, understanding failure modes, developing models for predicting failure modes, test the predictive model (s), and deploy for real-time operations as outlined above.

Even though there are many new sensors available for monitoring and control, there are many areas where manual steps are still required for maintenance activities due to lack of appropriate effective sensors. Thus the means of data acquisition can be multimodal. It can be obtained from samples collected manually and analyzed in the laboratory, monitored in real time with online sensors, acquired using portable data collectors, or examined by operators and engineers. In addition other tests and inspections at the machine site can help complete the picture and establish greater confidence in what's happening now (or not happening).The IIoT and AI technologies cannot make all other forms of condition monitoring obsolete, but they are powerful enablers. Consider an example in a machine shop with CNC routers, lathes, EDM machine, and metal 3D printers etc. Over a period of their operation the following can change: machine age which can lead to changing vibration, heat, acoustic emissions, displacement, alignment, balance etc.;oil and filter age; temperature and humidity; load conditions; looseness of parts; and operator handling. These changes may require further monitoring and adjusting such things as: oil flow rate and temperature control; grease dosage rate and frequency; viscosity correction; additive replenishment; machine operation; maintenance and inspection requisitions. Because of this complexity it is possible to design and develop different software applications that can address the above challenges by integrating more soft and hard automation processes.

AI Machine Learning involves the following steps: identifying the data set and corresponding sensors; collecting the data; preparing the data set for training, validation and testing; choosing a model and algorithm; perform model training calculations; evaluate the model; tune the model; and deploy the model for prediction.

### 4.1 CNC machine condition monitoring system

As mentioned above a major objective for manufacturing industries is to reduce the cost and improve safety and production. During 1980's to 1990's, the cutting tool was replaced based on wear of the cutting tool. But since then, tool condition monitoring systems have evolved and are used quite extensively to achieve the following objectives: early detection of cutting tool wear; maintaining a machining accuracy by providing a corrective action for tool wear; and prevention of cutting tool from breakage [18]. Using modern high precision sensors for sound, temperature and humidity, *vibration, strain/force, power,* and other appropriate *analog or digital sensors,* the user can monitor machining conditions using different software analysis options. Most of these common sensors provide a 0 to $\pm 10$ VDC analog signal, and 4 – 20 mA current signals. The data collected from these sensors can lead to limit analysis, spindle bearing faults, and frequency analysis etc.

In a recent study the authors used a MEMS installed sensor on a CNC machine with Fanuc controller to measure vibrations for maintenance purposes [19]. First, they carried out three case studies: optimal cutting values with a worn-out tool, cutting values with a new tool that breaks and does not break. After the analysis of case studies, a maintenance method was chosen according to TPM and TQM program guidelines. The analysis of the result lead to a proposed maintenance program. But the authors did not use any of the AI tools available to develop a predictive maintenance program.

### 4.2 Vibration detection application use case

In this case study a lathe is used to cut a stock bar into different size pieces. This lathe has an auto-fed bar feeder delivering 12-foot bar stock to the unattended running machine [20]. A significant number of parts were being scrapped due to irregularities in the bar, causing dimensional and finish errors. These irregularities in the bar could lead to a lot of scrap metal and damage the spindle. A commercial solution was used to install vibration sensor and connected it to the CNC control to monitor certain characteristics of the lathe. When excessive bar feeder vibration levels were detected the software would automatically signal the CNC to reduce spindle RPM until the vibration levels are acceptable to make good parts. If RPM must be reduced too much, and parts cannot be cut, an alarm was generated to inform the operator to stop the machine to remove the bar. In this case this process can be further automated, but it would require local technical resources to implement them.

### 4.3 Wireless monitoring

To monitor vibration measurement remotely the following protocol options are available: WirelessHART; ISA100; WiFi; Bluetooth; LoRa; and Proprietary. The following factors needs to be considered in selecting the appropriate protocol: built-in security; reliability; bandwidth; power consumption; supported configurations; interoperability between vendor products; and network maintenance. Another critical factor that needs to be considered is where do you perform the calculations related to the data analytics option for the wireless monitoring system and what information is going to be transmitted wirelessly. That is where and how would you view and store the raw sensor data, see the real-time trend and historical data, and send the alarms to mobile devices and remote computers. How and where would alarm be calculated.

WirelessHART (IEC 62591) is a field proven technology with very wide installed base as it was built on HART protocol. It has been an international standard (IEC

62591-1) since March 2010. This protocol offers the following advantages: it has small defined packet structure for reduced bandwidth and interoperability; built in security; uses mesh networking to ensure reliability; and can be expanded easily. A wide variety of device types are available from many automation equipment suppliers. SEPT Learning Factory provides the following two options for student projects: vibration transmitter (Emerson) as depicted in **Figure 5**; and a portable machine health analyzer (Emerson). The transmitter monitors/transmits WirlessHART vibration and temperature in hard-to-reach locations. It provides complete vibration information including overall levels, energy bands, high resolution spectra, and wave forms. It provides information for bearing and gear diagnostics. The PC hosts the specialized software provided by Emerson. The WirelessHART access point easily integrates into any host via Modbus TCP with capabilities for detailed diagnostics via a commercial software suite or custom-built Software. The wired option with dashboard shown in **Figure 6**, and user interface software components, **Figure 7**, required to build the dashboard for the Haas CNC vibration and other parameters monitoring has been implemented as well.

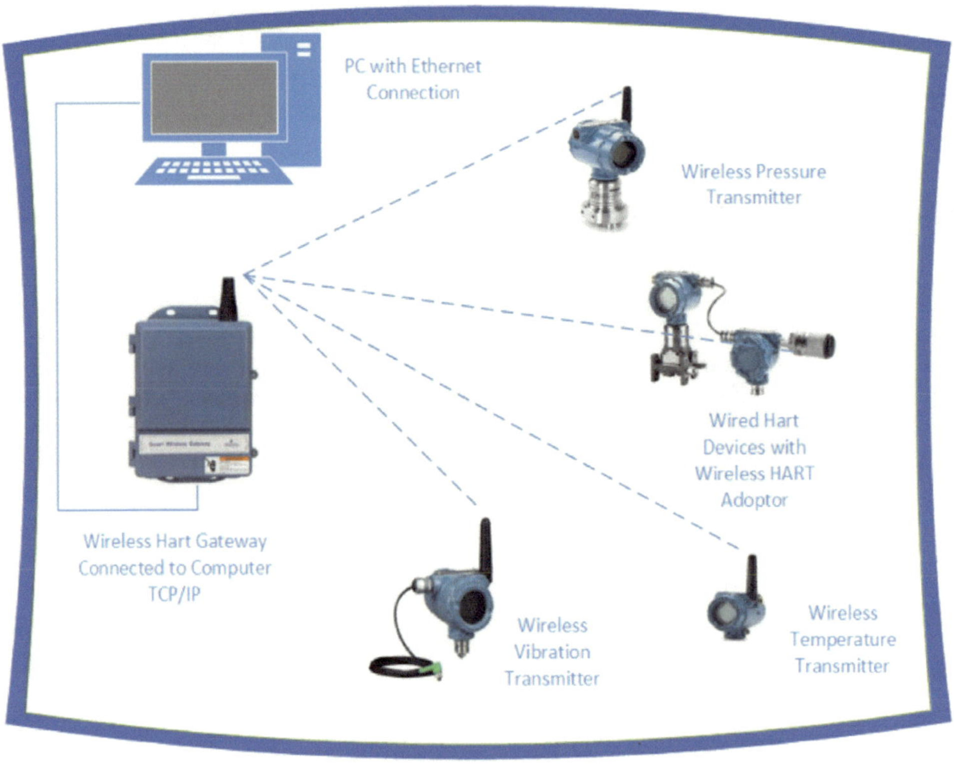


**Figure 5.**
*Wireless HART sample set up.*

On the other hand, the portable machinery health analyzer (Emerson) is used by the students working in the Learning Factory for vibration data collection and field analysis. This system can provide route vibration collection; advanced vibration analysis; cross-channel analysis; transient analysis; dynamic balancing; motor monitoring; and ODS modal analysis. It wirelessly uploads route data and corrective maintenance jobs from the field to AMS Machinery Health Manager (Emerson) for analysis and reporting. AMS Machinery Manager integrates data from multiple technologies, including vibration, oil analysis, thermography, and balancing into a

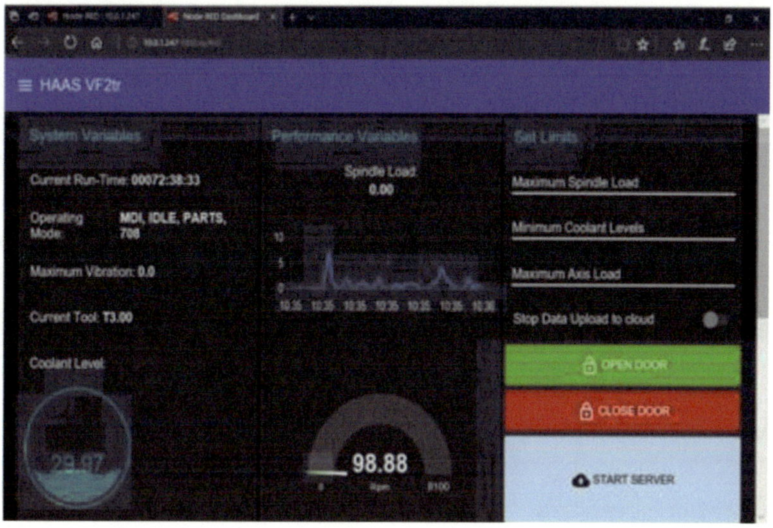


**Figure 6.**
*HAAS CNC web user Interface.*

single database to deliver the predictive intelligence necessary for increasing availability and reliability in the plant.

Another industrial IoT wireless machine health monitoring system with the following sensors: vibration sensor, thermocouple, AC split core current sensor, and ambient air temperature sensor is also available to the students for experimentation [21]. This low-cost device samples vibration, RMS current and temperature data and sends after a user-defined time interval over the wireless network. The vibration sensor samples 3-axis vibration data for 500 ms and then calculates RMS, Maximum, and Minimum vibration readings then combines these data with temperature values in a data packet and transmits the result to modems and gateways within wireless range. After each transmission it goes back to sleep, thus minimizing power consumption. This system uses DigiMesh® protocol, from Digi.com for wireless transmission of data, which automatically hops data from gateway to gateway until it arrives at the desired destination. The data on the other end is either received by an IoT gateway or an IoT modem, connected to local Learning Factory MQTT broker, and ultimately displayed on the dashboard designed by the students either using Grafana or Node-Red.

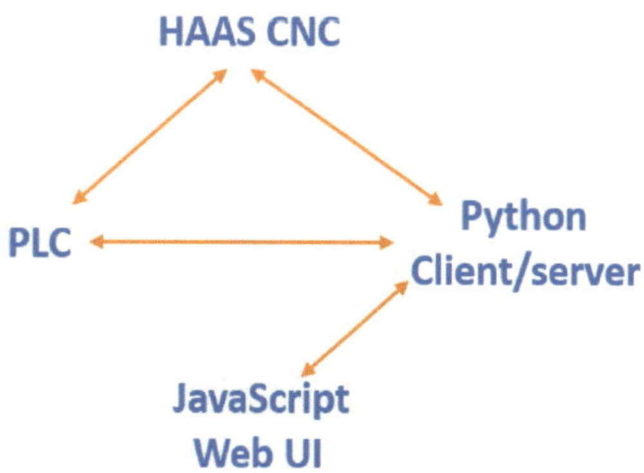


**Figure 7.**
*User Interface software components.*

It is also important to take note that the overall vibration data is not always a good indicator of machine health due to the following aspects of vibration measurement: fluctuates heavily due to process changes and insensitive to other failure modes such as bearing faults, gear defects, lubrication, and pump cavitation. Since the vibration analysis is based on raw vibration data other values such as RMS, peak value and impacting g's can be obtained as well. There are many approaches in interpreting the information from the raw vibration data. For example, contrary to claims in literature, it has been shown that RMS and peak values are good indicators of the gearbox health if used properly [22]. Another example is the analysis of vibration data from a process pump which showed that the overall vibration indicated good health of the asset while hidden in the raw data was the rising g values that indicated a bearing defect. The impacting analysis g values from the raw data can provide useful information for the following types of faults: bearing faults, gear defects, lubrication, and pump cavitation [23].

## 4.4 ISO vibration standards

20816-1 provides general guidelines;20816-2 covers land-based gas turbines, steam turbines and generators in excess of40 MW, with fluid-film bearings and rated speeds of 1 500 r/min, 1 800 r/min, 3 000 r/min and 3 600 r/min;10816-3 Industrial machines with nominal power above 15 kW and nominal speeds between120 r/min and 15 000 r/min when measured in situ; 7919-3 Mechanical vibration —Evaluation of machine vibration by measurements on rotating shafts—Coupled industrial machines; 20816-4 Gas turbines in excess of 3 MW, with fluid-film bearings; ISO 20816-5: Machine sets in hydraulic power generating and pumping plants,ISO 10816-6: Reciprocating machines with power ratings above 100 kW,Part 7: Rotodynamic pumps for industrial applications, including measurements on rotating shafts,Part 8: Reciprocating compressors.

ISO 20816-1 provides various evaluation zones: zone A, the vibration of newly commissioned machines normally falls within this zone; zone B, machines with vibration within this zone are normally considered acceptable for unrestricted long-term operation; zone C, machines with vibration within this zone are normally considered unsatisfactory for long-term continuous operation; zone D, vibration values within this zone are normally considered to be of sufficient severity to cause

| *mm/sec RMS* | *15 kW-300 kW Severity Zones* | |
|---|---|---|
| >11 | Zone D | Zone D |
| >7.1 | Zone D | Zone D |
| >4.5 | Zone D | Zone C |
| >3.5 | Zone C | Zone B |
| >2.8 | Zone C | Zone B |
| >2.3 | Zone B | Zone B |
| >1.4 | Zone B | Zone A |
| >.7 | Zone A | Zone A |
| >0 | Zone A | Zone A |
| Foundation | Rigid | Flexible |

**Table 1.**
*Criteria for assessing vibration measurements.*

damage to the machine. The **Table 1** provides summarizes this information for industrial machine falling with in the ISO 10816-3 standard.

ISO 10816-3:2009 is relevant to the common industrial machines and it gives criteria for assessing vibration measurements when made *in situ*. The criteria specified apply to machine sets having a power above 15 kW and operating speeds between 120 r/min and 15 000 r/min. Two conditions are used to classify the support assembly as shown in **Table 1**: rigid supports; and flexible supports. An important application of such a table is get a sense of what are the common measurement values used for the selection of sensor measurement range.

## 5. Machine health predictive maintenance teaching models

In this section we describe three machine health predictive maintenance teaching and training models developed at SEPT LF. They are: Fan Fault Detection and Diagnosis (FDD); IIoT Vibration Demonstration Station; and Machine Health Monitoring and Prediction Platform.

### 5.1 Fan FDD

#### *5.1.1 General overview*

The purpose of Fan FDD is to be able to determine faults in a mechanical system, in this case a spinning fan, and diagnose which part of the fan is faulty. It utilizes vibration and current measurements in combination with machine learning. The system uses a small computer called a Raspberry Pi, and a microcontroller (SAMD21G18A). Upon request from the Raspberry Pi, the microcontroller will record the vibration (in FFT) and current data from the fan and return it to the raspberry pi to store. The Raspberry Pi can use the data to train a neural network to make assumptions about the state of the fan.

The system is unique as it can be trained on 3 fault conditions, but it is able to output separate 4th condition which is a combination of two other conditions!

The circuit board contains a microcontroller, a SAMD21G18A, I2C for the accelerometer, ADXL345, an op-amp circuit for measuring current, and some display LEDs to give the user some feedback. The microcontroller can communicate with the Raspberry Pi through the Serial Port via USB. The firmware on the microcontroller only responds to commands from the Raspberry Pi serial port.

#### *5.1.2 Python programs*

The program that runs on the raspberry pi is written in python. Two main programs were written for Data Collection; and Fault Detection.

#### *5.1.3 Data collection program*

The purpose of this program is to collect different datasets to later train the network. The user will run through a series of events displayed on the screen. It will ask the user to turn on the fan, run the fan in normal condition, collect data, run the fan with a weight attached to the impeller (to simulate a broken fan blade), collect data, and finally run the fan with a cover on it (to simulate a blockage) and collect data. Once all of this has run it will automatically create a dataset and train the neural network with the data. The dataset for each fault condition is an FFT (calculated on the microcontroller) + the current measurement. Using the datasets, the data collection will run a program called "trainNN.py" which will output an neural network file type of "nn" which can be later used to determine the fault of the fan.

### *5.1.4 Fault detection program*

The purpose of this program is to use a previously trained neural network that used the data that was collected, and to determine the current state of the fan. The program will continuously run to give the state of the fan as an output on the raspberry pi. Since the training data collected was 3 individual fault conditions (including healthy) those are the states that the network will output. *The very interesting part of this system is, even though only three states were trained into the network, it can determine a 4th fault state which is the combination of both imbalanced impeller and a blockage at the same time!*

## 5.2 IIoT vibration demonstration station

With this vibration station the main purpose was to communicate the benefits of IIoT also known as "Industrial Internet of Things". At the time of development in 2016 this was a very new concept and so we thought we would showcase how it can be used in practice. Our system had a variety of components, starting with of course the mechanical/electrical but the bigger focus really was on the software front. In terms of software the system could be broken down into 3 major components, a frontend dashboard, a communications service for message passage, a data processing model to predict faults in the system.

### *5.2.1 Mechanical build*

For the mechanical build structure system, the goal was simple. We needed to create a structure to which we could mount the following components:

1. Motor with axle
2. PLC for motor control
3. Capacitive touch trigger
4. Power supply
5. DAC (Arduino used)
6. Accelerometer (sensor #1) *
7. DC current sensor (sensor #2) *

We also needed a method of being able to simulate a "failure event" the way we went about this was by creating an axel to which we could add counter weight to replicate the motion if a bearing had been torn due to overuse. By adding additional weight, we can create the level of counterbalance needed to test the flexibility of our predictive model later. Our result is as follows **Figure 8**.

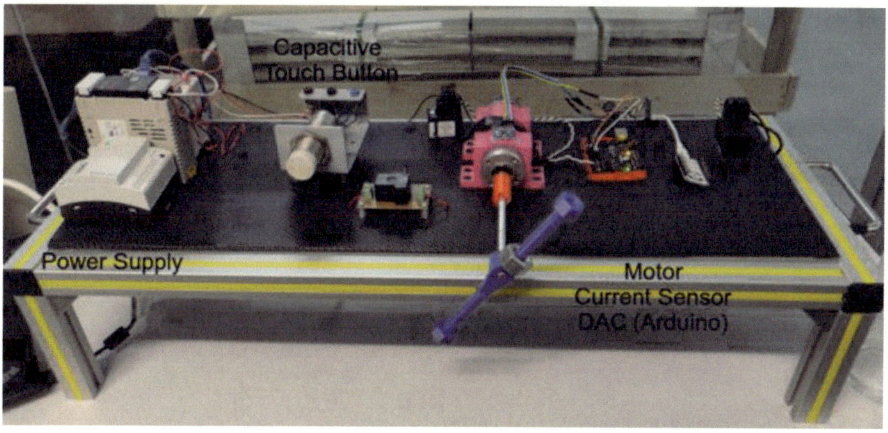


**Figure 8.**
*LF vibration demonstration.*

### *5.2.2 Electrical outline*

Our electrical system was also very straightforward, we simply needed to be able to provide the various currents required by all the system components. In the end we had the following devices actively connected in our system, in order of communication direction **Figure 9**.

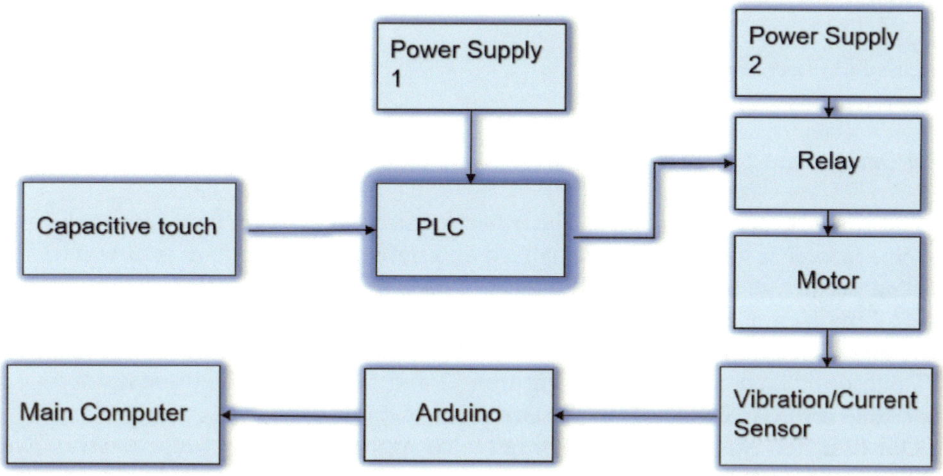


**Figure 9.**
*Block diagram of the electrical connections for the Vibration Station.*

### *5.2.3 Software visualization and data flow*

Our first objective for our IIoT vibration station was to get the data flowing and network so that we can visualize the raw information coming from the system. Our goal was to get the data front the sensor to a networked user interface that could be accessed anywhere from the world. This means we need a web app, data acquisition tool, and most importantly a messaging system. For our case we used MQTT, a middleware-based messaging system that allows for pub/sub based information transfer. In the end our data flow looked like follow **Figures 10** and **11**:

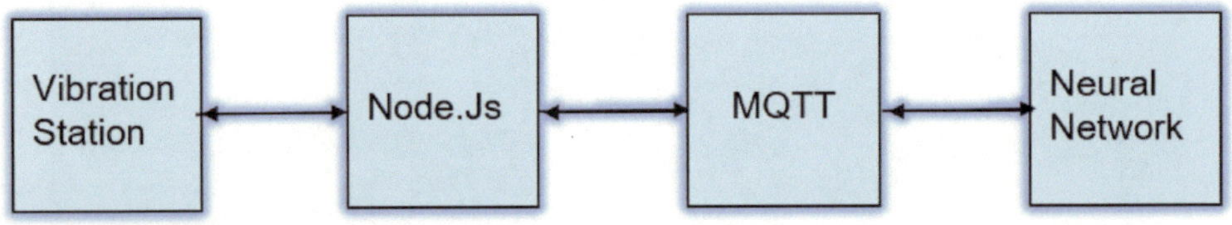


**Figure 10.**
*Vibration Station data flow components.*

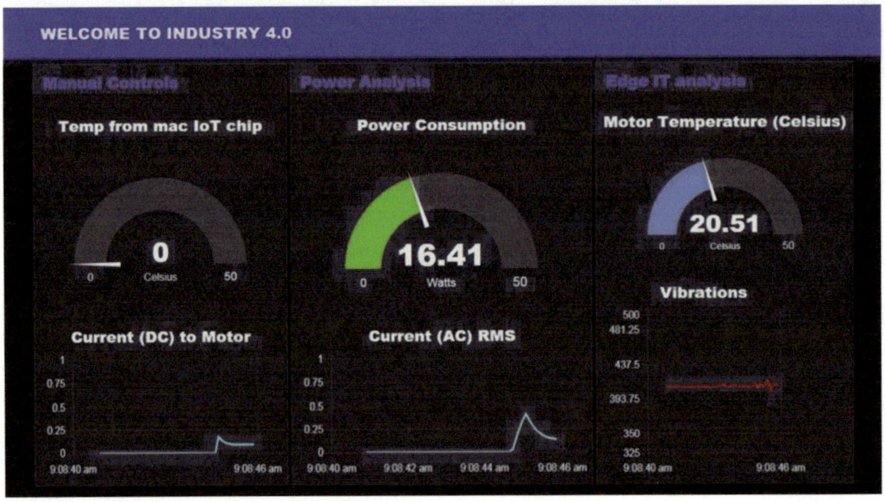


**Figure 11.**
*Vibration Station user Interface.*

### *5.2.4 Data science*

After having accomplished all the other requirements to get data from our system, as well as get that data visualized and networked, our main goal was to create a predictive maintenance model. We used our weighted offset assembly to create 2 main datasets, one where there is no imbalance and second where we add some offset weight and capture the "fault" labeled data. The difference between the FFT output of each can be seen in **Figures 12** and **13**, and as seen there is not a noticeable difference visually in the two plots.

Our first step was a vanilla neural network approach with no preprocessing, we fed the raw current and vibration data into a multi layered (3) neural network and trained it on both the good and bad data with the appropriate labels.

This approach turned out to be flawed, our main thinking in regards to this is that the neural network did not have enough learning capacity to map the given data into the frequency domain, where the faults are clearly indicated. The neural network would first need some sort of temporal mapping, and then on top of that also learn to differentiate between fault and no fault. To combat this issue, we attempted to preprocess the data through an FFT, this would map the data into the frequency domain and THEN through the neural network differentiate the data into it is given label.

This new approach worked much better, with a very high success rate in terms of predicting faults. The training also took only 5 minutes one a laptop, what this means is that the model can be trained on edge and in real time to continuously update itself as the system slowly degrades. Our simple neural network model is given as follows:

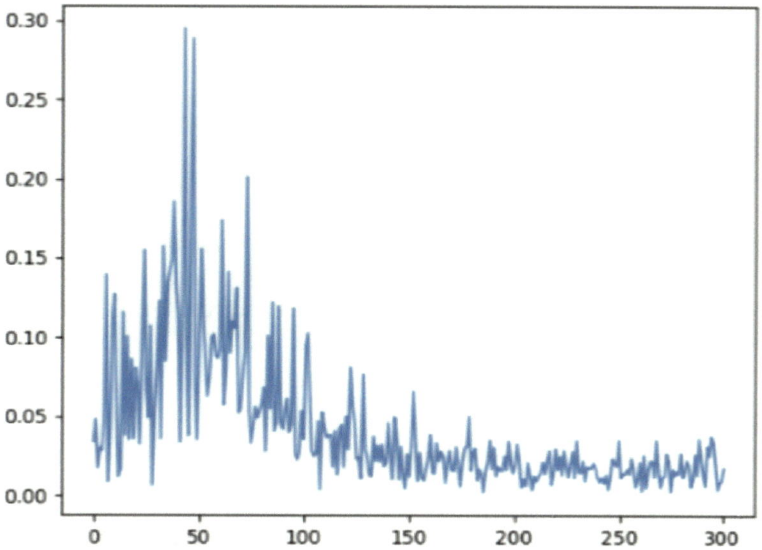


**Figure 12.**
*FFT plot with No load on the motor.*

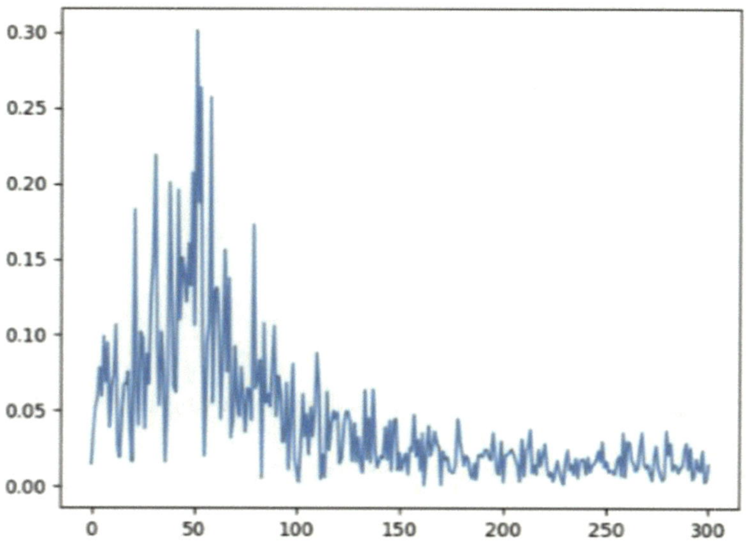


**Figure 13.**
*FFT plot with imbalance load on the motor.*

```
classifier = Sequential()
classifier.add(Dense(output_dim = 62, init = 'uniform',activation = 'relu',input_dim = 64))
classifier.add(Dense(output_dim = 1, init = 'uniform',activation = 'sigmoid'))
classifier.compile(optimizer= 'adam', loss = 'binary_crossentropy',metrics = ['accuracy'])
```

### *5.2.5 Conclusion*

An important takeaway from this project was around the structure of IIoT in a realistic setting as well as a realization of the various benefits IIoT brings, like the ability to visualize and see what is going on in our system in real time, from anywhere. This ability of constant perception into the system combined with a data processing model creates an approach for predictive maintenance that is very powerful, low-cost, and useful in active learning settings.

## 5.3 Machine health monitoring and prediction platform

An Advanced Predictive Learning Station has been developed, with an embedded vibration sensor and other sensors, that analyses a system for faults and transmits information wirelessly. Students will be able to report errors such as general imbalances, mechanical failure, resonance, electrical faults, and critical speeds. It can be easily extended to further application demonstrations such as bearing faults, AI modeling using multiple sensor inputs and analysis.

### *5.3.1 Lab setup and design*

The station design is based on an open architecture. It is designed to give students access to all potential avenues for learning and experimenting. The lab resource materials (designs, schematics, code) are all provided to the students. Documentation aside, students are also given access to as many elements on the learning station of the system as possible. There are oscilloscope probe points on the station to allow the students to view the trace of various signals as well as given control over many other aspects of the station. They can easily control the motor speed, explore all the signals of the system (even if they are not related to the lab). They can view the output of the encoder and view signal from the vibration sensor.

### *5.3.2 Station hardware layout*

The lab hardware is shown in a block diagram below. A detailed point by point description follows to express the high-level function of the components in the system (**Figure 14**).

### *5.3.3 Data access and signal analysis*

Because of the signal access points the students can using an oscilloscope to view the FFT plot on the screen. Most modern oscilloscopes have this feature built in. The students should clearly understand how the sampling system of the oscilloscope

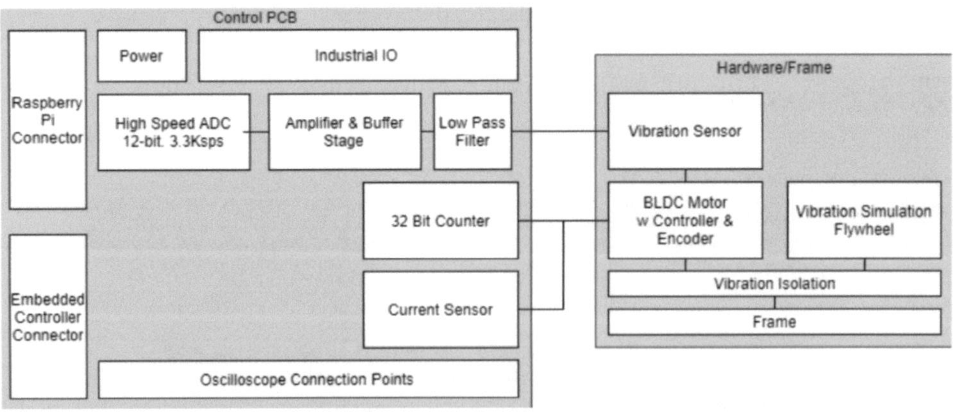


**Figure 14.**
*Machine health Hardware Station layout.*

relates to the standard parameters of an FFT. These should first be demonstrated by feeding a 1khz sine wave or square wave into the oscilloscope.

- **Sampling Rate:** In the below example the sampling rate is 1Ksps
- **Number of Bins:** The number of frequency bins the oscilloscope has
- **Number of Samples:** The number of samples used in each bin

In this example we are using 15000 samples at a sampling rate of 7.5 kbps on a 1 kHz sine wave. We are using 4096 frequency bins. From this experiment students will quickly see the application of an FFT in determining frequency components in complex time domain signals (**Figure 15**).

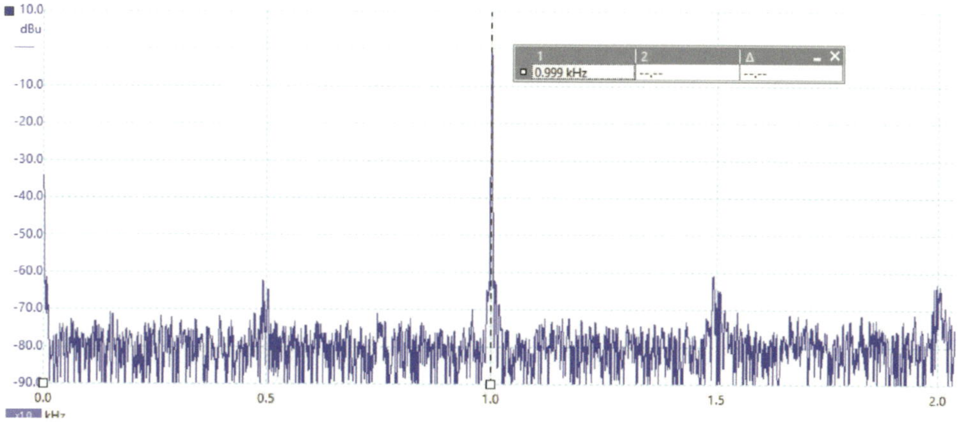


**Figure 15.**
*FFT plot of 1 kHz sinewave oscilloscope output.*

Once an understanding of the signal has been achieved, students can then do a lab experiment where they use the microcontroller to sample the incoming signal and do basic amplitude analysis on the signal. This can then be expanded to sampling on all three channels. A simple flow of sampling is demonstrated below. This sampling is done for single shot analysis, this is not a "real-time" sampling system. Realtime FFTs are typically done on FPGAs or systems with much more bandwidth capabilities (**Figure 16**).

The signal goes through an amplifier and buffer circuit to put it in the range of the ADC. The ADC is an I2C ADC, this is a better alternative to the micro-controller on board ADC because it is higher resolution. The microcontroller then samples the ADC at a known sampling rate (usually 2 kHz) and puts the samples into a buffer. This buffer can then be used for analysis.

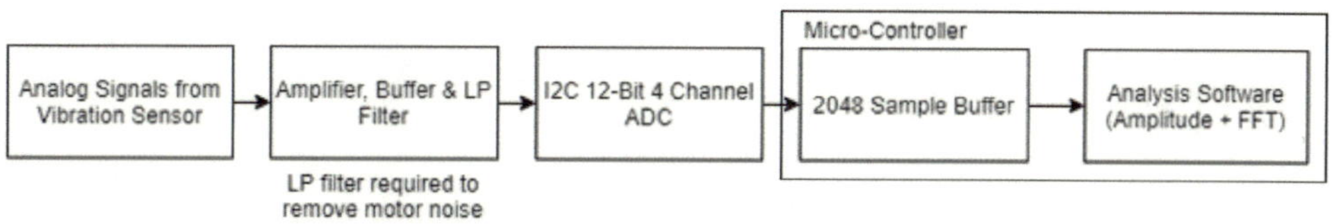


**Figure 16.**
*Flow of sampling data from the machine Health Station.*

If the students have a method of viewing the ADC buffer in a graphic way, it can help them develop the application faster. A good solution for this is using Python to receive serial/UART data from the micro-controller and plot it on a PC.

### *5.3.4 Wireless communication and data flow*

The final section of experimentation is designed to show students that for any data to be useful it must be provided to an end user who cares about it and it must be presented properly. The system of choice for communication is MQTT 5. This allows for the simplest possible method of publishing the resulting data from the microcontroller for external use (**Figure 17**).

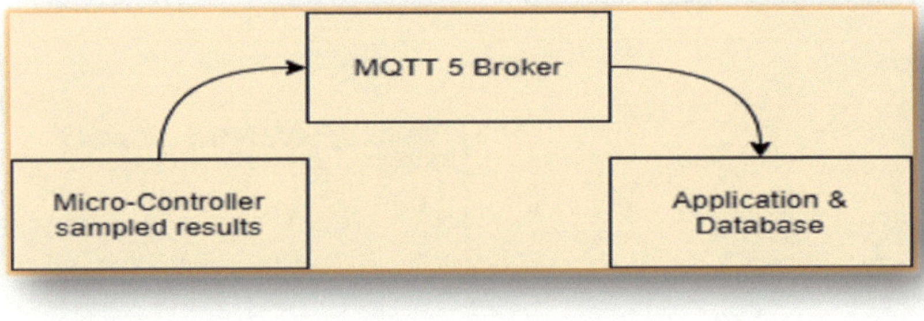


**Figure 17.**
*Machine health monitoring data flow.*

An InfluxDB can be used as the receiving database. It is important to display the data to the user and it is also important to store that data. If the data is stored it can later be used for developing machine learning based decisions. The historical data can also be used for long term vibration analysis which is the most important element in preventative maintenance.

This experiment can be extended to have students send the raw sampling data over Wi-Fi. What students will quickly realize (depending on their lab setup) is that

**Figure 18.**
*Machine health monitoring station hardware setup.*

it is much more "power expensive" to send raw data over Wi-Fi as compared to processing the data and sending the compressed details.

#### 5.3.5 Complete lab setup

The complete lab setup is featured below. The lab is designed to be stackable and easy to store. The main design feature of the system is the vibration isolation between the motor and the frame. If the motor is not correctly isolated, most of the vibration will be absorbed by the frame. This is conducive to real world implementation because motors are usually mounted on some type of vibration isolation barrier (**Figure 18**).

#### 5.3.6 Concluding remarks

This lab is designed to help students realize IoT and Industry 4.0 applications at an accelerated rate. There are some components of this application that were not explored. Students should be given brief readings to help them understand these additional components of vibration analysis.

- Variable speed drives will provide different vibration results at different RPMs; therefore, motor speed is reported.

- Varying loads on a motor (mixers, pumps, crushers) can provide data that is much more difficult to analyze due to the large volume assorted frequency components.

- Usually motor failures do not occur as one large change in a vibration peak but, rather, the slow increase of a particular peak that represents some wear or imbalance in the motor.

- A monitoring system should also consider system maintenance. If a component in the drive stage is replaced (a blade in a sawmill for example) the results will vary after the blade is change. There is no way to know the blade is changed without some user input or long-term data analysis. If there are students who work through this application at a faster rate, they can be asked to implement current monitoring as well.

## 6. Conclusion

The term IoT was coined in 1999 and it led to the evolution of IIoT enabled by other technologies that were being developed independently of each other. These technologies include such as: cyber-physical systems & cybersecurity; edge and cloud computing; mobile technologies; machine-to-machine communication; 3D printing; advanced robotics; big data; RFID technology; and AI. The SEPT Learning Factory is state-of-the art facility at McMaster University for the demonstration of integration of these technologies as well as development of educational/training material and resources, new technologies, and applied research.

Based on AI modeling the Fan Fault Detection and Diagnosis System described is able determine a fault state that was not included in training the AI model. The IIoT Vibration Demonstration Station using the neural networking model can detect the machine fault in real-time and publish the outcome on a dashboard connected via MQTT platform. The Machine Health Monitoring and Prediction Platform on the other hand combines the best features of the last two models for advanced predictive maintenance learning concepts and real-time demonstrations. This station has been developed, with an embedded vibration sensor and other sensors, that analyses a system for faults and transmits information wirelessly. Students will be able to report errors such as general imbalances, mechanical failure, resonance, electrical faults, and critical speeds. It can be easily extended to further application demonstrations such as bearing faults, AI modeling using multiple sensor inputs and analysis.

In this chapter we have described IIoT machine health monitoring foundations, models and applications for education and training. We have also illustrated how these models can be extended for development of predictive maintenance using AI technology.

## Acknowledgements

This project is partially funded by Future Skills Centre, Canada.

## Author details

Ishwar Singh[1*], Sean Hodgins[2], Anoop Gadhrri[1] and Reiner Schmidt[2]

1 SEPT, McMaster University, Hamilton, Canada

2 SEPT Alumnus, McMaster University, Hamilton, Canada

*Address all correspondence to: isingh@mcmaster.ca

## References

[1] Dalzochio, J. et al, Machine learning and reasoning for predictive maintenance in Industry 4.0: Current status and challenges, Computers in Industry, Volume 123, December 2020, 103298, doi:10.1016/j.compind.2020.103298.

[2] Gohel, H.A. et.al, Predictive maintenance architecture development for nuclear infrastructure using machine learning, Nuclear Engineering and Technology, Volume 52, Issue 7,2020, Pages 1436-1442, https://doi.org/10.1016/j.net.2019.12.029.

[3] Daniyan, I. et.al. ,Artificial intelligence for predictive maintenance in the railcar learning factories, Procedia Manufacturing, Volume 45,2020,Pages 13-18, https://doi.org/10.1016/j.promfg.2020.04.032.

[4] Einarson, G., How to Detect Pump Valve Failures in Positive Displacement Pumps: Case Study, 2021, https://reliabilityweb.com/articles/entry/how-to-detect-pump-valve-failures-in-positive-displacement-pumps-case-study

[5] Khanh, Nguyen & Medjaher, Kamal. (2019). A new dynamic predictive maintenance framework using deep learning for failure prognostics. Reliability Engineering & System Safety. 188. 10.1016/j.ress.2019.03.018.

[6] Gerasimos G. Samatas et.al, Predictive Maintenance –Bridging Artificial Intelligence and IoT https://arxiv.org/ftp/arxiv/papers/2103/2103.11148.pdf,

[7] Bump ,P. ,AI for Predictive Maintenance Applications in Industry – Examining 5 Use Cases, September 13, 2020, https://emerj.com/ai-sector-overviews/ai-for-predictive-maintenance-applications-in-industry-examining-5-use-cases/

[8] Puri, I. Tomorrow's engineers need to learn IoT, 2017, Available from https://www.design-engineering.com/features/tomorrows-engineers-iot/ [Accessed 2020-12-08]

[9] Mostafa S, and Singh I. Low Cost and User Friendly IoT Laboratory: Design and Implementation. In: ICBL2020 – 9th International Conference on Interactive Collaborative and Blended Learning: Springer, 2020.

[10] Centea D, Singh I, Elbestawi M. Framework for the Development of a Cyber-Physical Systems Learning Centre. In: BT - Online Engineering & Internet of Things., 2018.

[11] Elbestawi, M., Centea, D., Singh I, Wanyama, T. SEPT Learning Factory for Industry 4.0 Education and Applied Research. Procedia Manuf., 2018, Vol. 23.

[12] Top Six Vibration Problems. Available from https://houstondynamic.com/top-six-pump-vibration-problems/ [Accessed 2020-12-8]

[13] How the IIoT Is Changing Condition Monitoring Jim Fitch. Machinery Lubrication, April 2019, Available from https://www.machinerylubrication.com/Read/31496/iiot-condition-monitoring. [Accessed 2020-12-8]

[14] Machine Condition Monitoring Market by Monitoring Technique (Vibration Monitoring, Thermography, Oil Analysis, Corrosion Monitoring, Ultrasound Emission, MCA), Monitoring Process, Deployment, Offering, Industry, and Region - Global Forecast to 2025. MarketsandMarkets Research Private Ltd, Nov 2019. Report Code: SE 2739. Available from https://www.marketsandmarkets.com/Market-Reports/machine-health-monitoring-market-29627363.html [Accessed 2020-12-8]

[15] Singh I,Centea D, Elbestawi M. IoT, IIoT and Cyber-Physical Systems

Integration in the SEPT Learning Factory. Procedia Manufacturing, 2019, Vol. 31.

[16] Murphy, C. Choosing the Most Suitable Predictive Maintenance Sensor, Analog Devices. Available from: https://www.analog.com/en/technical-articles/choosing-the-most-suitable-predictive-maintenance-sensor.html. [Accessed 2020-12-8]

[17] Bryan,C. A Complete Guide To Predictive Maintenance. Available from https://limblecmms.com/blog/predictive-maintenance/.[Accessed 2020-12-8]

[18] Dimla, D. Sensor signals for tool-wear monitoring in metal cutting operations—a review of methods.: Int J Mach Tools Manuf, 2000, Vol. 40.

[19] Wszołek, G, et al Vibration monitoring of CNC machinery using MEMS sensor. Journal of Vibroengineering 2020, Vol. 22. Available from https://doi.org/10.21595/jve.2019.20788. [Accessed 2020-12-8]

[20] Barfeeder Vbration Success Story. Available from https://www.caroneng.com. [Accessed 2020-12-8]

[21] Industrial IoT Wireless Predictive Maintenance Sensor Available from https://store.ncd.io/product/industrial-iot-wireless-predictive-maintenance-sensor/. Accessed 2020-12-8]

[22] Igba, J. et.al. Analysing RMS and peak values of vibration signals for condition monitoring of wind turbine gearboxes. Renewable Energy, 2016.

[23] The Next Wave of the IIoT for Machine Monitoring. Available from https://reliabilityweb.com/videos/article/the-next-wave-of-the-iiot-for-machinery-monitoring. [Accessed 2020-12-8]